Nanotechnology For Dummies®

Cheat Sheet

What Is Nanotechnology?

Word	Definition
Nano	Greek for dwarf. Technically means "one billionth."
Nanometer (nm)	One-billionth of a meter; a common measurement unit of nanoscale science.
Nanotechnology	The National Nanotechnology Initiative defines nanotechnology as consisting of all of the following: • Research and technology development at the 1- to-100nm range. • Creating and using structures that have novel properties because of their small size. • Ability to control or manipulate at the atomic scale.

The Stuff Dreams Are Made Of

Illustration	Name	Definition	Applications
1 nm	Buckyball (or fullerene)	A soccer-ball-shaped molecule made of sixty carbon atoms.	Composite reinforcement Drug delivery
1.3 nm	Carbon nanotube	A sheet of graphite rolled into a tube.	Composite reinforcement Conductive wire Fuel cells High-resolution displays
5 nm	Quantum dot	A semiconductor nanocrystal whose electrons show discrete energy levels, much like an atom.	Medical imaging Energy-efficient light bulbs
100 nm	Nanoshell	A nanoparticle composed of a silica core surrounded by a gold coating.	Medical imaging Cancer therapy

For Dummies: Bestselling Book Series for Beginners

Nanotechnology For Dummies®

Tools of the Trade

- **Light:** Simultaneously exhibits properties of both waves and particles. As a wave, light has a frequency with a wavelength measured in nanometers. The particle is called a photon, an individual packet of light.

- **Raman spectroscopy:** Spectroscopic technique used to study vibrations and rotations of molecules in a sample.

- **Scanning electron microscope (SEM):** An electron beam shoots electrons at a sample. This process breaks off some of the sample's electrons, which are then read by a receiver. Provides 10nm resolution.

- **Transmission electron microscope (TEM):** An electron beam shoots through a sample and the electrons are read on the other side. The darker areas mean fewer electrons were transmitted through, indicating denser material; lighter areas mean more electrons transmitted through, meaning a less dense material. Provides 0.2nm resolution.

- **Atomic-force microscope (AFM):** A cantilever bounces up and down over the surface of a sample (similar to a needle on a music record) and maps out its contours.

- **Scanning tunneling microscope (STM):** A fine probe moves over the surface and a voltage is applied to both the probe and surface. Keeping the voltage constant, the probe moves up and down, providing atomic-size details of the surface. Provides 0.2nm or atomic resolution.

Computers and Electronics

- **Transistor:** Switches between the 1 (on) or 0 (off) bit in the computer processor.

- **Single-electron transistor (SET):** Similar to a regular transistor, but uses a single electron to indicate whether it represents a 1 or a 0. Using a single electron greatly reduces the energy required to run a processor, as well as limiting the heat levels produced during operation.

- **Magnetic random-access memory (MRAM):** Non-volatile electronic memory that uses less energy and is faster than conventional Dynamic RAM.

- **Spintronics:** "Spin-based electronics," which uses an electron's spin as well as its charge state to represent binary 1s and 0s.

- **Quantum computing:** A computer that uses quantum mechanical properties of superposition and entanglement to perform operations on data. Unlike conventional computers, a quantum computer will rely on probability (in effect, "it is highly likely that the answer is . . ."). The quantum computer will run in parallel, performing many operations at once.

- **Quantum cryptography:** Cryptography based on the traditional key-based cryptography, using the unique properties of quantum mechanics to provide a secure key exchange.

- **Photonic crystals:** Nanocrystals that guide photons (light) according to their structural properties. One possible application is as an efficient and practical optical router for use in Internet information exchange.

- **Microelectromechanical systems (MEMS):** Microscale machines with gears, motors, levers, and so on.

Wiley, the Wiley Publishing logo, For Dummies, the Dummies Man logo, the For Dummies Bestselling Book Series logo and all related trade dress are trademarks or registered trademarks of John Wiley & Sons, Inc. and/or its affiliates. All other trademarks are property of their respective owners.

For Dummies: Bestselling Book Series for Beginners

Nanotechnology

FOR

DUMMIES®

Nanotechnology FOR DUMMIES®

by Richard Booker and Earl Boysen

WILEY

Wiley Publishing, Inc.

Nanotechnology For Dummies®

Published by
Wiley Publishing, Inc.
111 River Street
Hoboken, NJ 07030-5774

www.wiley.com

For general information on our other products and services, please contact our Customer Care Department within the U.S. at 800-762-2974, outside the U.S. at 317-572-3993, or fax 317-572-4002.

For technical support, please visit www.wiley.com/techsupport.

Wiley also publishes its books in a variety of electronic formats. Some content that appears in print may not be available in electronic books.

Library of Congress Control Number: 2005924596

ISBN-13: 978-0-7645-8368-1

ISBN-10: 0-7645-8368-9

Manufactured in the United States of America

10 9 8 7 6 5 4 3

1O/RY/QX/QV/IN

WILEY

About the Authors

Richard Booker is a doctoral student at Rice University working under Dr. Richard Smalley, discoverer of the buckyball. He was lucky enough to get an outstanding education, starting with four intense years at Boston University earning a computer-engineering degree. After college, he joined the Air Force, where he managed weapon systems and researched battlespace networks while simultaneously working on his master's degree in computer engineering.

After four years, Captain Booker left the wild blue yonder to pursue his Ph.D. in applied physics at Rice and delve into the "new" world of nanotechnology. Rich's next ambitious adventure will be developing the armchair quantum wire (see Chapters 4 and 5 of this book) and helping to bring other nano-applications to market. His interests include flying, skydiving, scuba diving, music, art, movies and, in his spare time, working on being humble.

Earl Boysen is an engineer who, after 20 years in the computer-chip industry, decided to "slow down" and move to a quiet town in Washington. Earl is the co-author of *Electronics For Dummies,* and holds degrees in chemistry and physics. He lives in a house he designed and built himself, and keeps as busy as ever writing, acting, teaching math and science, dancing, and walking.

Dedication

For Him who gave us wisdom and heart to help each other.

For my fantastically supportive parents, Richard and Lorraine: Dad, whose insatiable appetite for knowledge, tenacity, and hard work ethic led me by example; and Mom, whose great teaching, organizational skills, and love kept me sane during my writing.

— R.B.

To my wonderful lady, Nancy, who (as well as providing support during the writing of this book) is the best thing that ever happened to me.

— E.B.

Authors' Acknowledgments

We'd like to thank our acquisitions editor, Katie Feltman, for giving us this opportunity and Richard Smalley for permitting his graduate student (Rich) to write this book in his "spare time."

Our project editor Paul Levesque and copy editor Barry Childs-Helton did a fantastic job at deciphering our nano-lingo and "Dummying" the rest. The graphic artists at Wiley did a great job at reproducing graphics — and a special thanks to those who provided use of photos free of charge; both fulfilled our vision of visually representing nanotech.

Special thanks to Nancy Stevenson for helping out on several of the chapters.

We'd also like to thank some of our tech reviewers:

Drs. Enrique Barrera, Wade Adams, and Howard Schmidt

Ph.D.s in training Joseph Cole, Aaron Saenz, Erik Haroz, Sungbae Lee, and Tushar Prasad

To our Web site developer, Jasyn Chen — thanks for all your hard work and generosity.

And yes, nanoscientists do marry supermodels.

Publisher's Acknowledgments

We're proud of this book; please send us your comments through our online registration form located at www.dummies.com/register/.

Some of the people who helped bring this book to market include the following:

Acquisitions, Editorial, and Media Development

Project Editor: Paul Levesque

Acquisitions Editor: Katie Feltman

Senior Copy Editor: Barry Childs-Helton

Technical Editor: Earl Boysen

Editorial Manager: Leah Cameron

Permissions Editor: Laura Moss

Media Development Manager:
Laura VanWinkle

Media Development Supervisor:
Richard Graves

Editorial Assistant: Amanda Foxworth

Cartoons: Rich Tennant (www.the5thwave.com)

Composition Services

Project Coordinator: Shannon Schiller

Layout and Graphics: Jonelle Burns, Karl Brandt, Carl Byers, Andrea Dahl Lauren Goddard, Stephanie D. Jumper Lynsey Osborn, Rashell Smith

Proofreaders: Leeann Harney, Jessica Kramer, Joe Niesen, TECHBOOKS Production Services

Indexer: TECHBOOKS Production Services

Publishing and Editorial for Technology Dummies

Richard Swadley, Vice President and Executive Group Publisher

Andy Cummings, Vice President and Publisher

Mary Bednarek, Executive Acquisitions Director

Mary C. Corder, Editorial Director

Publishing for Consumer Dummies

Diane Graves Steele, Vice President and Publisher

Joyce Pepple, Acquisitions Director

Composition Services

Gerry Fahey, Vice President of Production Services

Debbie Stailey, Director of Composition Services

Contents at a Glance

Table of Contents

Introduction

· ·

*H*ave you been reading the latest science magazines and noticed that the word "nano" has become as ubiquitous as the word "calorie" in a diet book? Are you looking for the latest hot investment opportunity and you've heard that something called *nanotechnology* will someday revolutionize our lives, and quite possibly your portfolio?

Well, you've come to the right place! *Nanotechnology For Dummies* is a fantastic way to get beyond all the hype and really understand what nanotechnology is and where it's headed. This is no dry, scientific journal; what you hold in your hands is a book that gives you just what you need to comprehend the basic concepts of nanotechnology, discover what nanotechnologists are making happen today, and get a look at the groundwork for tomorrow's nano-applications.

Why Buy This Book?

Even though nanotechnology deals with the very small, the topic itself is huge and complex. Like any technology, it consists of a lot of concepts and all sorts of cool but obscure jargon. Add to that, nanotechnology isn't just one science; it touches on a variety of sciences such as physics, chemistry, mechanical engineering, materials science, and biology. That's because, in all these disciplines and others, nanotechnology is simply the study of all things happening on a small — we mean REALLY small — scale that produce big (sometimes REALLY big) effects.

But this book provides you with just what you need to understand the basics of nanotechnology and its potential in your life and business in simple to understand language. This book doesn't pretend to answer all questions about nanotechnology because, frankly, nano is so cutting-edge that new questions about it are coming up every day. But what we've put together here does give you a good grounding in the essentials — and makes this exciting area of technology fun!

Why Nanotechnology?

We figure you have heard the term nanotechnology (unless you live in a cave) and have at least a passing interest in learning more, or you wouldn't have picked up this book. But what exactly is it about nanotechnology that makes it worth your while to explore?

Well, how about the probability that nanotechnology *is* the future? With the study of nano-size particles, devices, and composites, we will find ways to make stronger materials, detect diseases in the bloodstream, build extremely tiny machines, generate light and energy, and purify water.

In the business and investing world, the changes nanotechnology will bring in the next few decades will change the way people consume things — and, if you hadn't heard yet, we *do* live in a consumer society, which means that businesses and investors are more than slightly interested in technologies set to revolutionize the manufacturing of consumer goods. Fabrics in our clothing will become stronger and more stain-resistan — and keep us warmer. Cosmetics will gain skin-healing properties to keep us looking younger. TV flat-panel displays will become crisper, and solar cells will become more cost-effective. Drugs will have fewer side effects, as nanotechnology helps your body absorb them more completely and quickly.

In fact, there is probably not a walk of life that won't be affected by nanotechnology eventually. If that's not enough of a reason to read this book, how about this: Nanotechnology is just downright fascinating!

Foolish Assumptions

This book doesn't assume much, except that you have an interest in nanotechnology. You don't need a degree in physics or chemistry. We explain scientific concepts in a down-to-earth way so anyone with a passing interest in nanotech should have no trouble finding his or her way through the chapters. We also don't assume you know all the scientific buzzwords; that's why we define terms as we go.

We can't assume you will read this book from front to back in order. So if you decided to jump to Chapter 10 because it looks interesting (it is!), you won't be lost. We provide cross-references where relevant information is explained, and repeat the most important terms and ideas when necessary.

How This Book Is Organized

Nanotechnology For Dummies is organized so you can quickly find, read, and understand the information you need — and if you want to move around the various topics, you can skip chapters and move on to the parts that interest you.

The chapters in this book are divided into parts that help you get right to the information you're looking for. Here's the rundown . . .

Part I: Getting Small with Nanotechnology

Chapter 1 is where you get the overview of the nanotechnology realm, discovering what it is and how it is changing our world. You get an idea of how a nanotechnology product goes from concept to reality, as well as some of the challenges that have to be overcome to put nanotechnology in place in Chapter 2. Chapter 3 goes over those basic scientific concepts you should probably have under your belt if you really want to get your arms around nanotechnology, including what goes on inside molecules and how researchers are "seeing" things way too tiny to see.

Part II: Building a Better World with Nanomaterials

By building things from incredibly small particles, we can improve the strength of those materials. In Chapter 4 you hear all about things like carbon nanotubes, nanowires, and something called buckyballs. Chapter 5 examines how composite materials use nano-scale elements to make all kinds of things better, stronger, and faster — from plastics to bulletproof vests to who-knows-what-else. (Talk about limitless possibilities!)

Part III: "Smarter" Computers! Faster Internet! Cheaper Energy!

If Silicon Valley is your kind of place, you'll like this part. Chapter 6 is where we look at all the things being done with transistors, computer-chip manufacturing, and computer memory. In Chapter 7 we tell you how people are telling photons where to go, using light to improve telecommunications and other industries. Chapter 8 covers how nanotechnology is changing electronics with new ways to generate light, sense things, and build tiny electromechanical devices. This stuff is so cool that . . . well, let's just say you won't find it at your local Radio Shack (yet)! Finally, Chapter 9 looks at how cleaner energy can be generated with nanotechnology, and how it might improve our environment in the future.

Part IV: Living Healthier Lives

Chapters 10 and 11 are where you hear all about the promise of nanotechnology in diagnosing and curing all that ails us. From more efficient drug delivery to sending tiny diagnostic machines into our bloodstreams, to vastly improving medical imaging and the mapping of our genes, medical science is super-excited about nano-scale solutions.

Part V: Investing in Nanotech

Corporations, universities, and military/government labs are all working together to further nanotechnology research because it's got so much promise in so many applications. Discover in these three chapters where the research is happening, where advances are being made, and where the big wins are likely to occur. If you're interested in speculating on the future of nano with your checkbook, this part is for you.

Part VI: The Part of Tens

Every *For Dummies* book rounds things out with a few chapters that follow that tried-and-true List of Ten format. In Chapter 15 you hear all about ten movers and shakers in nanotechnology, and in Chapter 16 we cover ten resources in print and on the Web that you just have to check out if we've managed to get you excited about the nano realm.

Icons Used in This Book

A picture is worth a thousand words, so this book uses little graphic icons to visually point out useful information that you may want to know more about.

The Tip icon indicates information that might be of interest for further exploration or lead you into new ideas. These icons tend to point out tidbits that make exploring nanotechnology more enjoyable, so don't just skip 'em!

Remember icons are gentle reminders about important ideas or facts that you should keep in mind while exploring nanotechnology. We also use these icons to cross-reference places in the book where we talk about a topic in more detail, so you can flip to those chapters and brush up if you want to.

We can't help it: Occasionally there will be technical information in this book. (No kidding?) But we thought you might appreciate a little warning when the relatively obscure stuff crops up. So when there's highly technical information just ahead (even if it's all really interesting stuff, as any of us completely unbiased authors would tell you), we may tuck one of these icons next to it.

Going Online

This book has its very own Web site, www.nanotechnologyfordummies.com. This is your go-to source for updated information on the fast-changing world of nanotechnology. Go to this Web site to ask us questions and get updates on nanotechnology.

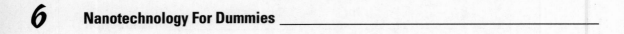

Part I
Getting Small with Nanotechnology

In this part...

Nanotechnology is the hot buzzword of the new millennium. But what, exactly, is it?

In this part, we start sketching out the answers. We give you the broad overview of nanotechnology in Chapter 1 — what it is, where it came from, what's being produced today, and where the whole thing is heading. We address some of the hurdles that nanotech will have to jump to make progress (Chapter 2), and Chapter 3 gives you a look at some scientific terms, concepts, and souped-up nano-equipment needed to delve deeper into the topic.

Chapter 1

The Hitchhiker's Guide to Nanotechnology

In This Chapter

▶ Finding out what nanotechnology is and how it will change your world

▶ Identifying the difference between real science and science fiction

▶ Investing wisely in the emerging nanotech industry — and still keeping your shirt

. . . necessity . . . is the mother of our invention.

from The Republic *by Plato (c. 370 B.C.)*

Welcome to the world of nanotechnology — technology capable of fulfilling our every need (almost). It's safe to assume that you know a little bit about nanotechnology from picking up this book. However, you may have a few unanswered questions. Maybe you've heard it described as "The Next Industrial Revolution" on the news followed by some business commentary. Maybe you're a Will Smith fan and saw his 2004 movie *I, Robot,* where "nanites" save the day, dismantling the main computer from the inside. Other than a financial topic or clever plot device, what is nanotechnology (exactly)? Do I *need* nanotechnology? Will I be able to cash in on a "nanotechnology bubble" or will I lose my shirt as I did with the dot-coms? These are all fair questions; we address each of them in this chapter.

Grasping the Essence of Nanotechnology

We start off this chapter by defining nanotechnology and showing you not only the need but also the inevitability of this technology. We then go into detail explaining what you can expect from nanotechnology. (Short, sharp, and to the point — that's our motto!)

Finding out what it is

Nano, Greek for "dwarf," means one billionth. Measurement at this level is in *nanometers* (abbreviated "nm") — billionths of a meter. To put this into perspective, a strand of human hair is roughly 75,000 nm across. On the flipside of the concept, you'd need ten hydrogen atoms lined up end-to-end to make up 1 nm. Figure 1-1 illustrates the differences in scale that range from you all the way down to one hydrogen atom.

The definition

Nanotechnology can be difficult to determine and define. For example, the realm of nanoscience is not new; chemists will tell you they've been doing nanoscience for hundreds of years. Stained-glass windows found in medieval churches contain different-size gold nanoparticles incorporated into the glass — the specific size of the particles creating orange, purple, red, or greenish colors. Einstein, as part of his doctoral dissertation, calculated the size of a sugar molecule as one nanometer. Loosely considered, both the medieval glass workers and Einstein were nanoscientists. What's new about current nanoscience is its aggressive focus on developing applied technology — and the emergence of the right tools for the job.

When faced with a squishy term that can mean different things to different people, the best thing to do is to form a committee and charge it with drawing up a working definition. In fact, a committee *was* formed (the National Nanotechnology Initiative) and the following defining features of nanotechnology were hammered out:

1. Nanotechnology involves research and technology development at the 1nm-to-100nm range.

2. Nanotechnology creates and uses structures that have novel properties because of their small size.

3. Nanotechnology builds on the ability to control or manipulate at the atomic scale.

Numbers 1 and 3 are pretty straightforward, but Number 2 uses the eyebrow-raising term "novel properties." When we go nano, the interactions and physics between atoms display exotic properties that they don't at larger scales. "How exotic?" you ask? Well, at this level atoms leave the realm of classical physical properties behind, and venture into the world of quantum mechanics. David Rotman described this best in his 1999 article, "Will the Real Nanotech Please Stand Up?" (published in the March/April edition of *Technology Review*), when he quoted Mark Reed, a nanoelectronics scientist at Yale University:

> "Physical intuition fails miserably in the nanoworld . . . you see all kinds of unusual effects." For example, even our everyday electrons act unusual at the nano level: "It's like throwing a tennis ball at a garage door and having the ball pop out the other side."

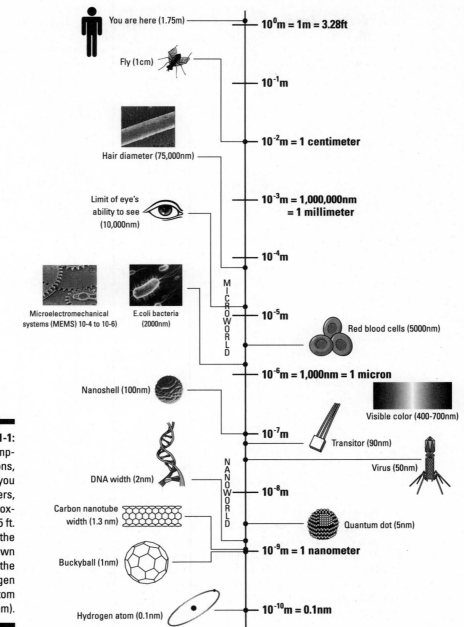

Figure 1-1:
Size comparisons, from you (1.75 meters, or approximately 5 ft. 7 in.) all the way down to the hydrogen atom (0.1 nm).

Need another concrete example? Check this one out. It is demonstrably true that a gold nanoparticle has a color, melting point, and chemical property different from those you'd find in a macro-scale Fort Knox gold brick. That's because the interactions of the gold atoms in the larger gold brick average out — changing the overall properties and appearance of the object. A single gold nanoparticle, on the other hand, can be its own idiosyncratic self — a tiny object, free from the averaging effects of countless other gold atoms.

The applications

Nanotechnology is, at heart, interdisciplinary. You'll get only part of the story if you just use chemistry to get at the properties of atoms on the nano level — adding physics and quantum mechanics to the mix gives you a truer picture. Chemists, physicists, and medical doctors are working alongside engineers, biologists, and computer scientists to determine the applications, direction, and development of nanotechnology — in essence, nanotechnology is many disciplines building upon one another. Industries such as materials manufacturing, computer manufacturing, and healthcare will all contribute, meaning that all will benefit — both directly from nanotechnological advances, and indirectly from advances made by fellow players in the nano field. (Imagine, for example, quantum computers simulating the effectiveness of new nano-based medicines.)

There are two approaches to fabricating at the nano scale: top-down and bottom-up. A *top-down* approach is similar to a sculptor cutting away at a block of marble — we first work at a large scale and then cut away until we have our nano-scale product. (The computer industry uses this approach when creating their microprocessors.) The other approach is *bottom-up* manufacturing, which entails building our product one atom at a time. This can be time-consuming, so a so-called *self-assembly* process is employed — under specific conditions, the atoms and molecules spontaneously arrange themselves into the final product. (Self-assembly is described further in Chapter 8.)

Some science-fiction plots — they know who they are — revolve around this self-assembly concept, conjuring up plot lines infested with tiny self-replicating machines running amok. (For a closer look at this far-fetched notion, see the "Welcome to Nano Park" sidebar in this chapter.) For the near-term, it looks like the top-down approach will be favored because it tends to provide us with greater control (and, more importantly, it uses some time-tested techniques of the computer industry). If we were betting men — which we are not, because as men of science we know that the House always wins — we would venture that the top-down approach will be the fabrication method of choice for quite awhile.

Evolving into Nanotech

If you take a look at the world around us, you'll notice that nature herself designs at the molecular level. Nanotechnology intends to imitate nature by taking advantage of the unique properties of nano-scale matter to come up with more efficient ways of controlling and manipulating molecules. With technology, smaller is better. If you take a look at technical evolution, you'll notice that we're continually getting smaller — computers the size of a room in the 1950s now fit on your lap; cellphones the size of a brick in the 1980s now fit in your shirt pocket. Consumer convenience, the economics of resources and competition, and the advantages of faster processing, higher productivity, and better quality all play a part in motivating companies to go small.

Technology is not only getting smaller, it's also evolving faster. As new technologies develop, we build upon previous knowledge. Thanks to the Internet, this knowledge base — and rate of information exchange — is increasing rapidly. To illustrate this evolution of technology, check out the evolution of electrical components: from the vacuum tube to the solid-state transistor to the carbon nanotube field effect transistor.

✔ **Vacuum Tube (1897):** This was the ancestor of the transistor, essentially a light bulb with three, instead of two, terminals. It was large, hot, and prone to burning out.

✔ **Solid-State Transistor (1947):** Instead of using a filament, the solid-state transistor switches between On and Off using different materials — metals and semiconductors. This let us come up with transistors that were smaller in size, didn't give off as much heat, and were far more durable.

✔ **Carbon Nanotube Transistor (1998):** A carbon nanotube (refer to Figure 1-1) — a graphite sheet rolled into a tube — comes in two main forms, metallic and semiconducting. The carbon nanotube was discovered in 1991 and within only seven years, was used for shuttling electrons across two electrodes. Not only is it incredibly small (nano-scale), but it also uses less energy and gives off less heat by using few electrons to indicate whether it's on or off. (For more on carbon nanotubes, see Chapter 4.)

It took 50 years to get from the vacuum tube to the first solid-state transistor — and another 50 years of refinement to get solid-state transistors to be all they could be. But when we'd developed the needed tools and understanding, it only took seven years from the discovery of the carbon nanotube to turn it into what may be the ultimate transistor.

The history

The word "nanotechnology" proper was coined by Nario Taniguchi in 1974 to describe machining with tolerances of less than a micron. But this really isn't the term's true beginning. Three noteworthy events and discoveries got this ball rolling — all by Nobel laureates (and when a Nobel Prize winner says something, you listen . . . and try to understand).

✔ **The Vision:** In 1959, Caltech physicist Richard Feynman gives his famed talk "There's Plenty of Room at the Bottom," outlining the prospects for atomic engineering. (To read the talk in its entirety, check out www.its.caltech.edu/~feynman/plenty.html for a handy transcript.)

✔ **Seeing is Believing:** In 1981, Gerd Binnig and Heinrich Rohrer of IBM's Zurich Research Laboratory create the scanning tunneling microscope, enabling researchers to both see and manipulate atoms for the first time. (For more on the scanning tunneling microscope, see Chapter 3.)

✔ **Nanostructures:** In 1985, Robert F. Curl Jr., Harold W. Kroto, and Richard E. Smalley discover buckminsterfullerenes (buckyballs — refer to Figure 1-1), soccer-ball-shaped molecules made of carbon and measuring roughly 0.7nm wide. (For more on buckyballs, see Chapter 4.)

Why you want nanotechnology in your life

Nanotechnology will increase your standard of living — no ifs, ands, or buts. Done right, it will make our lives more secure, improve healthcare delivery, and optimize our use of limited resources. Pretty basic stuff, in other words. Mankind has spent millennia trying to fill these needs, because it has always known that these are the things it needs to ensure a future for itself. If nanotechnological applications pan out the way we think they will pan out, we are one step closer to ensuring that future.

Security

Security is a broad field, covering everything from the security of our borders to the security of our infrastructure to the security of our computer networks. Here's our take on how nanotechnology will revolutionize the whole security field:

✔ **Superior, lightweight materials:** Imagine materials ten times stronger than steel at a fraction of the weight. With such materials, nanotechnology could revolutionize tanks, airframes, spacecraft, skyscrapers, bridges, and body armor, providing unprecedented protection. Composite nanomaterials may one day lead to shape-shifting wings instead of the mechanical flaps on current designs. Kevlar, the backbone fiber of bulletproof vests, will be replaced with materials that not only provide better protection but store energy and monitor the health status of our soldiers. A taste of what's to come: MIT was awarded a $50 million Army contract in 2002 to launch the Institute for Soldier Nanotechnologies (ISN) developing artificial muscles, biowarfare sensors, and communications systems.

✔ **Advanced computing:** More powerful and smaller computers will encrypt our data and provide round-the-clock security. Quantum cryptography — cryptography that utilizes the unique properties of quantum mechanics — will provide unbreakable security for businesses, government, and military. These same quantum mechanics will be used to construct quantum computers capable of breaking current encryption techniques (a needed advantage in the war against terror). Additionally, quantum computers provide better simulations to predict natural disasters and pattern recognition to make *biometrics* — identification based on personal features such as face recognition — possible.

✔ **Increased situational awareness:** Chemical sensors based on nanotechnology will be incredibly sensitive — capable, in fact, of pinpointing a single molecule out of billions. These sensors will be cheap and disposable, forewarning us of airport-security breaches or anthrax-laced letters. These sensors will eventually take to the air on military unmanned aerial vehicles (UAVs), not only sensing chemicals but also providing incredible photo resolutions. These photos, condensed and on an energy-efficient, high resolution, wristwatch-sized display, will find their way to the soldier, providing incredible real-time situational awareness at the place needed most: the front lines.

✔ **Powerful munitions:** Nanometals, nano-sized particles of metal such as nanoaluminum, are more chemically reactive because of their small size and greater surface area. Varying the size of these nanometals in munitions allows us to control the explosion, minimizing collateral damage. Incorporating nanometals into bombs and propellants increases the speed of released energy with fewer raw materials consumed — more (and better-directed) "bang" for your buck.

Healthcare

Making the world around us more secure is one thing, but how about making the world *inside* us more secure? With nanotechnology, what's beneath our skin is going to be more accessible to us than it's ever been before. Here's what we see happening:

✔ **Diagnostics:** Hospitals will benefit greatly from nanotechnology with faster, cheaper diagnostic equipment. The lab-on-a-chip is waiting in the wings to analyze a patient's ailments in an instant, providing point-of-care testing and drug application, thus taking out a lot of the diagnostic guesswork that has plagued healthcare up to now. New contrast agents will float through the bloodstream, lighting up problems such as tumors with incredible accuracy. Not only will nanotechnology make diagnostic tests better, but it will also make them more portable, providing time-sensitive diagnostics out in the field on ambulances. Newborn children will have their DNA quickly mapped, pointing out future potential problems, allowing us to curtail disease before it takes hold.

✔ **Novel drugs:** Nanotechnology will aid in the delivery of just the right amount of medicine to the exact spots of the body that need it most. Nanoshells, approximately 100nm in diameter, will float through the body, attaching only to cancer cells. When excited by a laser beam, the nanoshells will give off heat — in effect, cooking the tumor and destroying it. Nanotechnology will create biocompatible joint replacements and artery stents that will last the life of the patient instead of having to be replaced every few years.

Resources

The only thing not in short supply these days is more human beings — and we're not about to see a shortage of them any time soon. If we are going to survive at all — much less thrive — we are going to need to find ways to use the riches of this world more efficiently. Here's how nanotechnology could help:

✔ **Energy:** Nanotechnology is set to provide new methods to effectively utilize our current energy resources while also presenting new alternatives. Cars will have lighter and stronger engine blocks and frames and will use new additives making fuel more efficient. House lighting will use quantum dots — nanocrystals 5nm across — in order to transform electricity into light instead of wasting away into heat. Solar cells will finally become cost effective and hydrogen fuel cells will get a boost from nano-materials and nanocomposites. Our Holy Grail will be a reusable catalyst that quickly breaks down water in the presence of sunlight, making that long-wished-for hydrogen economy realistic. That catalyst, whatever it is, will be constructed with nanotechnology.

✔ **Water:** Nanotechnology will provide efficient water purification techniques, allowing third-world countries access to clean water. When we satisfy our energy requirements, desalinization of water from our oceans will not only provide enough water to drink but also enough to water our crops.

You say you want a revolution?

We predict that a few revolutions will roll over us in the course of the next 50 years — revolutions that will have great impact on our lifestyles and economy, and will involve nanotechnology, energy, and robotics. We've broken down each revolution here, complete with an estimated peak year for each one's public and financial popularity. Note, however, that nobody's jumped in a time machine to check up ahead; these peak years are based on the "gut feelings" of the authors, and on our research and observations of current trends. But we aren't just spinning tales. In February 2005, Business Week Online polled its readers, asking when they thought nanotechnology would change their lives: 34 percent said by 2007; an additional 51 percent said by 2015. At this point, it's not a matter of "if" but a matter of "when" — it's nearly as certain as death and taxes.

✔ **Nanotechnology itself:** We're thinking **2012** is about the time that significant revolutionary products will be available, along with solid companies within the industry. A "nanotechnology bubble" will begin to develop around 2010, but this may not be as drastic as the "Internet bubble." (See the "Getting a (Small) Piece of Nanotechnology for Yourself" section, later in this chapter.) Low-power, high-density computer memory, longer-lasting batteries, and some medical applications (including cancer therapy and diagnostics) will be some of the early products (pre-2010).

Advances in computer processing will follow (2015), and new materials and composites will come online toward 2020. The order of events will likely be computers and medical first, and then materials — all overlapping but peaking at their respective years.

Energy: Our crystal ball says **2025** is going to be the Nanotechnology Energy Year. With demand for energy rising in industrializing countries such as China and India, oil will continue to be that highly-sought-after resource. Oil prices and prosperity have an inverse relationship — as oil prices go up, prosperity goes down. Goldman Sachs has recently (April 2005) suggested that we may enter into a "super-spike" period of oil demand, with prices as high as $105 a barrel — almost twice the current price. It has also been suggested (by Princeton University geologist Kenneth Deffeyes) that world oil production will peak around Thanksgiving 2005. Unfortunately, production capacity has grown more slowly than demand, which makes things even worse.

Nanotechnology will combine efficient use of our current sources while providing directions to explore for alternate sources of energy. Nanomaterials that emerge around 2020 will not only provide lighter/ stronger materials for vehicles but will also improve efficiency in the collection, storage, and transmission of energy, greatly aiding our transition from gas to solar, hydrogen, or maybe even renewable bio-fuels (for example, vegetable oils and bioalcohols such as ethanol and methanol).

Robotics: Think **2045.** This may seem a little farfetched, because you expect something with the personality of C3-PO and the powers of a Jedi, but today you're getting R2-D2 — just bells and whistles. However, there are a few driving forces making robotics economically feasible: defense, space exploration, and labor. In the near term, autonomous UAVs (short for *U*nmanned *A*erial *V*ehicles) will keep continuous watch over our borders, and robots will dispose of roadside bombs in the battlefield. Space exploration will be done by robots, cutting the need for human involvement and thus allowing us to go farther than we've ever gone before.

As nanotechnology develops better sensors and processors and the energy revolution provides abundantly cheap energy, robots will be in demand as cheap manual labor, increasing our overall standard of living. Not only will we have robotic dogs and vacuum cleaners but also assembly-line industrial labor, bringing money back to Western nations. Perhaps a "robotic arms race" will emerge, not as a mighty military machine but as a productivity machine — each nation trying to make the cheapest goods as quickly as possible. All this will gradually grow over the next few decades — but once the hardware is in place (around 2030), the software and artificial intelligence will soon follow.

Some of these years may seem a long way off, but these changes will arrive faster than you may think. If you're currently in college, they'll happen in your lifetime.

Knowing what to expect (and not expect)

Futuristic excitement aside, our expectations for nanotechnology need to be realistic and we need to be patient, for not all the advances that nanotechnology is set to bring will happen overnight. Nanotechnology will not be a miracle cure. Although there will be some fantastic advances, not everything that we imagine will come to fruition. However, nanotechnology is also sure to usher in things that we never envisioned coming — products that could end up changing the world.

Nano-scale science isn't a free-for-all — there are rules. We won't be able to manufacture something that, at the molecular level, is chemically unstable. Scientists know how most things work chemically and physically, but there have been a few surprises — and we learn the rules along the way. Time to take a look at some examples of the nanotechnology we now have, what we can improve upon, what will be new, and what (we can confidently say) will never happen.

What we have

Nano applications are already showing up in areas as diverse as computing, transportation safety, and medicine. The steps may seem modest by future standards, but they get big effects from tiny things. Three examples illustrate what we can do now:

✔ Computer transistors have broken below the 100nm barrier — transistors are officially nano-sized. Look for the devices that house them to shrink as well, and for devices that are already small (such as cellphones) to become more powerful.

✔ Airbag sensors, although micro in size and bigger than nano, are used in most recent cars — some of them already saving lives. These sensors will continue to shrink, becoming more powerful and accurate.

✔ At least one home pregnancy test (Carter-Wallace's "First Response") uses both gold nanoparticles and micrometer-size latex particles on an external, disposable test sheet. The product takes advantage of how gold nanoparticles of different sizes reflect light differently. If a woman is pregnant, a specific hormone is present that causes the micro-sized and nano-sized particles to clump together — and those bigger particles reflect a distinctive color: a visible pink strip appears on the test sheet. If she's not pregnant, no hormone is present, which means no clumping of the nanoparticles and no pink strip.

What will be improved

Besides introducing new products and procedures, nanotech will advance those that already exist. In January 2005, the Lemelson-MIT Program identified the top 25 innovations of the past 25 years. Nanotechnology was number 21 . . .

which is good. What's even better is that the other 24 would all be positively influenced by nanotechnology. Try these examples on for size:

- ✔ Cellphones with longer battery life
- ✔ Global Positioning Systems that are smaller and more accurate
- ✔ Computers that are faster and smaller
- ✔ Memory storage that packs greater capacity into a smaller space and uses less energy
- ✔ DNA fingerprinting that is quick and accurate

Other items not on the list will be oil additives designed to get more out of our precious resource, new medical diagnostics and drug delivery . . . even an aesthetically pleasing sunscreen. (Most sunscreens are a white, thick and sticky cream. Nanophase Technologies has developed a sunscreen that is transparent — the active ingredient is a nano-scale material that, because of its small size, doesn't scatter visible light.) Nanotech will crop up everywhere in existing products, even — or especially — in places you can't see.

What will be new

Nanotechnology promises to be a cornucopia of wonders — improving our healthcare, optimizing our use of resources, increasing our standard of living. Detecting disease at the molecular level will lead to new treatments for old ills. The development of materials ten times stronger than steel — but a tenth of the weight — offers to make transportation faster and more efficient. (Imagine, for example, how air transportation would change if airframes were lighter and stronger, plane engines used less fuel, and sensors and smart material automatically deformed the wings to minimize drag.) New, nanotech-based paints and coatings will prevent dirt and water from adhering to surfaces such as kitchen counters, vinyl siding, cars, and windows. (Imagine a car you never have to wash, that rolls dirt and water right off so you don't even have to use your windshield wipers.)

Speaking of cleanliness, EnviroSystems' EcoTru disinfectant cleaner is the only EPA-registered Tox Category IV disinfectant product — it doesn't harm the skin, eyes, lungs, or body if ingested. Conventional disinfectants dissolve in a solvent, and are meant to drown the organisms with toxic chemicals — which can be about as bad for humans as for small organisms. EcoTru uses nanospheres of charged oil droplets suspended in water to carry the active ingredient that ends up penetrating the microorganisms' membranes. This stuff is so good that Doctors Without Borders used EcoTru in the operating room as an antiseptic when they ran out of their regular antiseptic. Of the 500 patients that they used EcoTru on, none got an infection. EcoTru is also already used as a disinfectant on airplanes, on cruise ships, and in healthcare facilities, and, given Doctors Without Borders's experience, may be used as an antiseptic in the future.

What will not happen

Science fiction writers describe swarms of molecule-size robots swimming through your bloodstream cleaning your arteries while shooting cancer cells. And the nanobots that aren't fixing your body are out there fixing and building the world around us, one atom at a time. These scenarios are highly improbable, if not impossible. If they do eventually prove possible, they're decades (if not centuries) away.

As marvelous as it is to envision nanobots curing our bodies and quickly assembling and disassembling inanimate objects, these methods may not even be the most efficient approach. After all, some of the best medicine involves coaxing the body to help fix itself — and building inanimate objects one atom at a time (even something as simple as a chair) is no quick task. There may be a better, more inventive way to use engineering principles at the nano scale — one that takes advantage of the opportunities that chemistry and intermolecular interactions offer. But those opportunities are far more modest and (well, yeah) small-scale than science fiction would suggest.

The dramatic creation and transformation of macro-scale objects makes for spectacular entertainment but dicey science. Here's why: Molecular chemistry is very complex and involves controlling atoms in three dimensions. At each reaction site, the atoms feel the influence of neighboring atoms. To do any mechanics at this level (which is what nanobots would have to do for this to work), you would need to control the motion of each and every atom — a very difficult juggling act.

A lot of this nano-zealous science fiction got started in 1986, when K. Eric Drexler, founder of the Foresight Institute (a nonprofit organization dedicated to educating the public about nanotechnology), penned *Engines of Creation: The Coming Era of Nanotechnology*. In it, he describes self-replicating nanoassemblers building objects one atom at a time. He also describes a doomsday scenario referred to as "gray goo" — myriads of self-replicating nanoassemblers making uncountable copies of themselves and consuming the earth. "Gray" because they're machines; "goo" because they're so small they'd look like a thick liquid. Scientists have since ridiculed this Drexlerian vision — even Drexler himself (in 2004) said "runaway replication" was unlikely.

In the December 1, 2003, edition of *Chemical & Engineering News,* Eric Drexler and Richard Smalley, Nobel Prize winner and discoverer of the buckyball, squared off, arguing for and against molecular manufacturing. The "Point-Counterpoint" article was a series of letters between the two, where Drexler continues to outline his "mechanical" molecular manufacturing, whereas Smalley argues against such a model by describing a need for "chemistry" even at the nano level. One source of contention is the "gray goo" scenario. Drexler had presented this scenario as a warning to not let nanotech get out of hand whereas Smalley sees it as unnecessarily scaring the public on a doomsday scenario that's a) highly unlikely and b) threatens public support for nanotech by harping the negative. In the end, they both disagreed on molecular manufacturing but continue to promote nanotechnology's potential.

Welcome to Nano Park

In 2003, Michael Crichton, the author of *Jurassic Park*, published his next sci-fi doomsday book, *Prey*. Crichton takes us to the top-secret research labs of Xymos Technology, where the self-replicating microscopic machines prey on the scientists. In the end, the humans prevail (sorry to ruin the ending, but hey, *somebody* had to be left alive to buy the book). This book, and others like it in the grand tradition of high-tech disaster fiction, pilfers the most visually stunning aspects of an emerging technology and then presents the worst-case scenario of what could go wrong. Without a conflict (nanobots running amok) there's no resolution (humans winning) and no story. Unfortunately, in an effort to make his story plausible, Crichton mixes a little too much science fiction with reality — even the company name Xymos suggests the name of a real nanotech company (Zyvex). Here are some of the many flaws:

✔ Crichton confuses two basic tools of nanotechnology, the scanning-probe microscope and an electron microscope (see Chapter 3 for an explanation of both).

✔ The description of a nanobot is a bit off — "one ten-billionth of an inch in length." This is the size of a single atom — certainly not of a whole robot.

✔ The nanobots are too big to fit inside a synapse and control human beings. Synapses, junctions interconnecting the neurons of the central nervous system, are only a few atoms wide and these nanobots are at least a hundred times bigger. This claim ties in with the previous one — nanobots "one ten-billionth of an inch in length." Although Crichton is consistent in his size reference, nanobots composed of only a few atoms wouldn't have the computational power to do anything that the book claims they can do.

These books may offer some benefit by generating interest in the subject and drawing people into science. Unfortunately, they may also instill some unnecessary fear and anxiety in our society as new technologies emerge.

Getting a (Small) Piece of Nanotechnology for Yourself

Nanotechnology is set to insinuate its way into our economy in ways that we can only imagine — more probably, in ways that we could never dream. Nanotechnology as an industry will be hard to identify and track, considering it's very pervasive — it touches upon a lot of different industries. Familiar products will be the first to take advantage of nanotech — clothing, cosmetics, and novel industrial coatings. When the processes to make these products are mastered, new and innovative products will begin to emerge. Businesses will rise, and those that don't adapt to this changing environment will fall by the wayside. Now that you're salivating at the prospects of making a quick buck, eager to enter the Lilliputian world of atoms and molecules, this section of the chapter will temper your appetite and present what we believe to be a *realistic* picture of the current industry — and how it will probably play out.

The nanotech industry

The National Science Foundation projects that nanotech will be a $1 trillion industry by 2015 — that's 10 percent of the current GDP of the United States. Small wonder (so to speak) that the National Nanotechnology Initiative has increased federal funding for nanotechnology research and development from $464 million in 2001 to $982 million in 2005 (as shown in Figure 1-2). Governments across the world poured $4 billion into nanotech research in 2004.

However, according to Lux Research Inc., only $13 billion worth of manufactured goods will incorporate nanotechnologies this year. In the grand scheme, that's not very much. Nanotechnology, because of its complexity and reach over different industries, will grow slowly over many decades. Therefore, we must be patient and not expect a huge explosion of products coming online all at once.

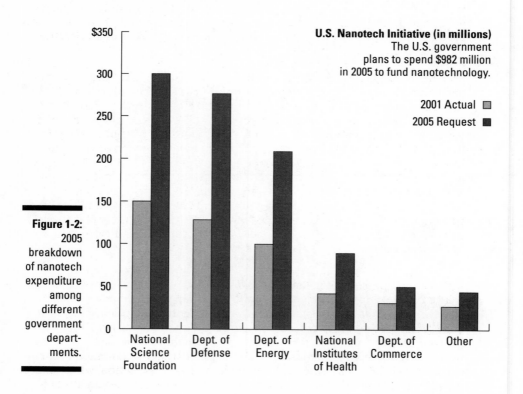

Figure 1-2: 2005 breakdown of nanotech expenditure among different government departments.

U.S. Nanotech Initiative (in millions)
The U.S. government plans to spend $982 million in 2005 to fund nanotechnology.

2001 Actual
2005 Request

Today, nanotech companies are either developing their knowledge base through research or are producing the materials for other people's nanotech research. There are three entities from which nanotechnology will emerge: open research (universities and national labs), large corporations, and start-ups. These entities, in turn, have three options: They can produce their discoveries, license them, or sell the rights to them outright. Licensing may be an attractive option — it creates cash flow with minimal overhead.

Big companies will have some advantage here, given their resources and ability to purchase high-end measurement equipment. Some large industries, such as pharmaceuticals and microchips, will be able to successfully integrate nanotechnology because they already have the processes and facilities in place to get their product to market — something small companies may not have access to.

However, this shouldn't discourage small companies who have the flexibility to adapt. Small companies are set to develop products, processes, and intellectual property, becoming attractive takeover candidates by big companies. In the end, everyone wins — investors make money, small companies develop products rapidly and efficiently, and big companies produce and distribute the end product to the consumer.

Both large and small companies, given nanotech's reach, will have to develop partnerships and collaborations — not only between different industries but also between different companies, universities, and government labs. Not only will they exchange research but also resources. For example, Rice University has a partnership with the Texas Medical Center, the largest medical center in the world. Additionally, Rice (as well as other universities across the nation) has broken down some financial barriers by developing the Shared Equipment Authority (SEA), which will train and allow businesses to use million-dollar measurement equipment for reasonable prices. If you want (for example) to learn how to use a scanning electron microscope, it will cost you $200 to get trained and $20/hour for each subsequent use. That's incredibly cheap, considering that the equipment costs $500,000 to begin with, and more than $40,000/year just to keep it maintained.

Battle of the bubbles: Nanotech versus Internet

Every industry has an *economic bubble* — speculation in a commodity that causes the price to increase, which produces more speculation, which causes the price to increase . . . until, at some point, the price reaches an absurd level and the bubble bursts, causing a sudden — and precipitous — drop. Bubbles aren't good — they encourage people to misallocate resources in ways that don't pan out, with nonproductive results. Additionally, the crash that follows can cause great economic problems. Of course, if you don't get carried away, an economic bubble isn't all bad — it allows money to flow into a new industry to give it financial support to grow from. But too much blind enthusiasm too soon can backfire as market forces weed out ventures that don't produce actual profits in a reasonable time.

Many industries have had bubbles — the railroad industry, the automobile industry, even the tulip industry had an economic bubble. The most recent example was the Internet bubble in the late 1990s. Alan Greenspan, chairman of the Federal Reserve Board, identified this emerging bubble in 1996, calling investor speculation "irrational exuberance," which should have been enough to warn off most folks — but fools and their money are soon parted.

So, is it foolish to invest now in nanotechnology? Not necessarily. We have no doubt that there will be a "nanotechnology bubble" (within the next five years?), but we hope it will prove to be more "exuberant" and less "irrational." One thing to keep in mind is that, with the dot-com bubble, companies were started for $5,000 by lawyers and marketing agents who were pushing an idea — and not necessarily a product. Nobody was quite sure what to do with the idea yet. The essentials of the scientific knowledge base could be pretty well absorbed by your average information-technology technician after only a few months of training. Nanotechnology and nanoscience research, on the other hand, are a much more complex kettle of fish. They require in-depth scientific knowledge and a Ph.D.-level background. This knowledge base will be indispensable to any nanotech start-up company, ensuring that the field will not be inundated by every Tom, Dick, and Harry.

This requirement of high-level technical know-how is slowing nanotechnology's growth. There is a shortage of talent — a limited supply of scientists coming up against increasing demand. Additionally, product-cycle times — the time from research to market — is long. On the flipside, nanotech has fewer barriers to market adoption than the Internet had. In order for online stores to be successful, customers had to own a computer, establish an Internet connection, and gain confidence in online transactions. Nanotechnology will already be integrated into existing products — no massive new-product-adoption process will be required.

Nanotechnology's biggest advantage over the Internet as The Next Big (Little?) Thing will be its patentable intellectual property. Web-based innovations were difficult to patent, allowing competitors to quickly adapt and clone a product within months. Nanotechnology's intellectual property creates a major barrier to entry by being incredibly difficult to replicate. The time cost to replicate is measured in years, encouraging the competitor to either take a different approach or license the patent. This barrier to entry is fantastic for small companies; they have more time to develop without getting immediately crushed by the big dogs.

What's in a name?

Nanosoft . . . Nanosonic . . . Nano Inside. These are just a few words and phrases that folks have tried to trademark with the United States Patent and Trademark Office. Whether they'll be successful, we don't know yet. "Nano" has recently become the generic word for "high tech" — its cachet is similar to what "virtual" and "holo" had in the past few decades.

A few companies that already have "nano" in their names do loosely develop some "nano" products. U.S. Global Aerospace, for example, which uses nanofiber technology in its cockpit doors, recently changed its name to U.S. Global

*Nano*space — that's a drop of 14 orders of magnitude! Nanogen makes *micro* arrays for genetic research. Its NanoChip array has test sites 80 microns in size, spaced 200 microns apart — not quite nano but headed in the right direction.

These quasi-legit nano-namings may have a hidden bonus — they tend to make investors leery, prompting them to scrutinize the claims and dig deeper into a company's nano-legitimacy. After all, we shouldn't be able to "just add nano" to sell a stock or company. But don't be surprised if some folks try that anyway.

Caveat Emptor — Buyer Beware

If recent history is any indication, our coming nanotech bubble may be overhyped and fueled with stock speculation. In this section, we hope to quell some of this speculation early on and paint a realistic picture for you. Currently, very few nanotech companies are public (that is, offering stock for public purchase) — and the ones that are have a stock chart that reads more like a cardiogram than a steady line of growth. That hasn't deterred some early players; on the Dow Jones Industrial Index, 19 of the 30 companies have launched nano initiatives. This section takes a look at two particular companies and throws in some indexes you can follow as well.

Nanosys

In April 2004, Nanosys filed for a $115 million initial public offering (IPO) with the Securities and Exchange Commission (SEC). They are a company rich in patents — over 200 — but alas, no profits yet (sigh). This is not to say they won't ever make money — they are one of the companies that our research shows has some great potential — and it's a pretty sure bet they'll license their research.

For now, they're cautious. In the Nanosys SEC filing, the company stated: "To date, we have not successfully developed any commercially available products. . . . We do not anticipate that our first products will be commercially available for at least several years, if at all." (*If at all?* Not the rosiest picture.) Nanosys even withdrew its filing in August 2004, stating "volatility of capital markets." But Nanosys has clout as one of the poster-child nanotech companies — and we give them credit for not starting a wild nanotech bubble before the industry has had time to mature.

Altair Nanotechnologies, Inc.

On February 10, 2005, Altair Nanotechnologies, Inc. [NASDAQ: ALTI], announced a breakthrough in their lithium-ion battery-electrode material. This novel nanomaterial — composed of nano-size lithium-titanium-oxide particles — does the following for their rechargeable batteries:

- **Delivers more power:** You get three times the power of existing lithium-ion batteries, to be exact.

- **Allows faster recharge:** Recharge takes a few minutes instead of hours.

- **Ensures longer life:** The material improves the number of recharge and discharge cycles from a few hundred to many thousand cycles.

On this news, the stock spiked. Here's what the results looked like:

- **Day before:** Closed at 2.08 with volume of 500k.

- **Day of:** Closed at 4.77 (129.3% increase) with volume of 56.6 million.

- **Day after:** Peaked at 6.52 (another 36.6% increase) with volume of 101.6 million.

Not bad for two days' worth of work. As of mid-April, 2005, ALTI had a 50-day moving average of 3.8 with an average volume close to 10 million. On April 19, 2005, they announced the initial shipment of their electrode nanomaterial for testing at a partner company, Advanced Battery Technologies, Inc. This nanomaterial will be incorporated into Advanced Battery's Polymer-Lithium-Ion (PLI) batteries tested in electric vehicles.

Investment tools and strategies

Whether or not Altair Nanotechnologies or Nanosys are going to be the next eBay is certainly up in the air. Given the volatility of owning individual stocks in nanotechnology, it's wise to go with something a little more diverse, such as an exchange-traded fund (ETF) or mutual fund. Unfortunately, there are no nanotech ETFs or mutual funds at this time. Harris & Harris [NASDAQ: TINY] is as close as you're going to get for now; it's a publicly traded venture-capital firm specializing in nanotech companies.

Merrill Lynch established a "Nanotech Index" [Amex: NNZ] in 2004 — on April Fools Day, of all days (those guys are such kidders). The index consists of 22 small companies whose future business strategy is based on nanotechnology. Lux Research, Inc., is a research and advisory firm focusing on the economic impact of nanotech businesses. Their index, Lux Nanotech Index [Amex: LUXNI], follows 26 publicly traded nanotech companies. Merrill Lynch's Nanotech Index is open for trading, but Lux Nanotech Index is not.

The world according to nano

Nanotechnology, given its scientific complexity, is going to require a large amount of upfront capital — and substantial government assistance. In 2003, 52 percent of the $5.5 billion invested in nanotechnology came from national governments. The following table shows the Venture Analytics breakdown of nanotechnology investment.

Nano Investment

Country	2003 Nanotech Funding in Millions of Dollars	Percent from National Government
Japan	1,610	50
U.S.	1,524	51
China	480	58
South Korea	280	71
Germany	218	54
Australia	193	48
U.K.	160	56
World Total	5,544	52

This chart can be misleading — a wide gap in financing does not necessarily result in a wide gap in manpower. Although the United States and Japan outspent China, engineers and scientists in China make between one-sixth and one-tenth of what Americans earn. The United States spends five times as much as China but has less than half as many researchers (1.3 million Chinese versus 734,000 American). Additionally, China's universities and vocational schools produced 325,000 engineers this year — five times as many as the United States.

As you conquer your greed and quell your fears (both wise moves), here are a few generic tips to keep in mind. For further financial understanding, there's a copy of *Investing For Dummies* (authored by the inestimable Eric Tyson and published by Wiley) with your name on it.

- **Follow the money:** There will undoubtedly be huge investments into nanotech. Be mindful of large volume shifts, indicating large institution investment.

- **Follow the trend:** 75 percent of all stocks follow the Dow Jones Industrial Average — this will include nanotech stocks. If stocks are going up, buy; if going down, sell. Buy low, sell high.

✔ **Don't listen to tipsters:** Tipsters will, more than likely, be wrong about a lot of nanotech. We still need those scientists to help business analysts make sense of what's plausible and realistic. Do your own research; don't let a news sound bite influence your decision and control your money.

✔ **Set a stop loss:** Set a sell-point with your brokerage firm; make it roughly 7 percent below your purchase price — and as the price goes up, increase this stop-loss point. Doing so prevents what happened to many when they rode the 1990s Internet crash into the ground.

Chapter 2

Nano in Your Life

*I*f you read Chapter 1, you know that nanotechnology is the great hope of the future, a better mousetrap, a Holy Grail of technology that could make things stronger, people live longer, and stuff in general just great. That's true, to an extent. But before we delve deeper into the wonders of nanotechnology, it's only fair to look at the challenges that have to be met.

What challenges? First, it's not a piece of cake getting from a nano-concept to a viable nano-product in the marketplace. Second, there are ethical concerns — just as there are with any new technology — that could pose barriers to nano-progress. Finally, there is a global concern that we could create a nano-divide, with rich nations benefiting from nano-progress and poorer nations lagging behind.

Going from Lab to Factory to Home

Here's something you can bank on: In the future, we are going to see numerous new applications based (to some degree) on nanotechnological advances. However, going from an idea to a laboratory to a product that appears in your kitchen or bathroom can take years (or decades) and lots of work. Take a look at what it took to transform a lab idea about a molecular-level product into something called Kevlar.

What's a Kevlar?

Kevlar was developed by DuPont back in the 1960s — a particularly turbulent time in U.S. history. It turns out that DuPont had a track record in nylon and other fibers that gave them an incentive to look for even better fibers. At some point, they set a goal for themselves of creating a fiber with super-heat-resistant properties (like asbestos, but without the health risks and lawsuits) and with a stiffness almost like glass. Kevlar was the answer: It's made up of something called an aramid fiber, and it has about five times the strength of steel. You might find Kevlar in products such as bulletproof vests, fire-blocking fabrics, cables used in a whole bunch of applications, or even materials for reinforcing tires or airplane fuselages.

When Kevlar was developed, in the pre-nanomania 1960s, there were a great many hurdles DuPont had to jump — and it had to call on lots of disciplines to get that job done. At times, current thinking had to be circumvented in order to move forward. No surprise that an environment that encouraged questions and challenges to the status quo was a key to the Kevlar success story.

When you look at it phase by phase, this project is a great example of what it means to bring molecular-level products from concept to market.

Phase One: Research

It was 1965. A research scientist at DuPont's Pioneering Research Laboratory named Stephanie Kwolek was hard at work trying to make a lightweight fiber that was strong and could withstand high temperatures. She had discovered that she could fit together certain molecules (namely, para-aminobenzoic acid) into a long, chainlike molecule — a polymer. The problem was that the thinking of the day denied that any opaque polymer solution (such as this one) could be woven into fiber. Stephanie decided to keep going anyway. She tried extruding the polymer through spinneret, a piece of equipment that worked like a spider spinning silk, except in this case it took polymers in solution and spun them into fiber.

But is it nano?

Kevlar was developed over 40 years ago. It's not often referred to as a nano-discovery, because few people spoke about nano then. When this work was done it was called polymer chemistry, even though it used nano-size materials.

Even today, you have to remember that nanotechnology is merely the application of anything on a molecular level, in any field. One researcher may refer to a work in progress as a chemistry or physics project; another may refer to the same endeavor as nanotechnology. Some people hold that 10 or 20 years from now there will be no nanotechnology per se, just the use of nano-size materials in a variety of disciplines. It's sort of like the way people use colorings in lots of things: foods, clothing, car paint, and so on. There's not exactly a field of colorings, just the use of them in a variety of industries.

To everybody's surprise, the opaque polymer spun quite well, and made a kind of super-fiber. In fact, its stress-strain curve — a standard measurement of fiber strength — was startling, so startling that the lab had to test the results over and over before anybody believed what they were seeing. Almost as an added bonus, the heat resistance was just what they were looking for.

The bad news was that the raw material was very expensive. To find a cheaper alternative, a huge program was launched to try to understand the physical chemistry of this type of polymer. In the process, they found something called PPD-T, a suitably similar type of polymer made from lower-priced ingredients.

But that wasn't the end of their problems. It turned out that in order to put the cheaper polymer through the spinneret, they had to dissolve it in sulfuric acid, something that tends to burn a hole through people. In addition, the sulfuric acid mixed with the polymer was so thick (*viscous* to all you chemist types out there) that it wasn't practical to get the spinneret up to the required speed to get it to spin into fiber.

The researchers weren't about to give up, so they went to the manufacturing-and-engineering groups. These farsighted engineers essentially told them to go jump in the nearest lake. The spinning solvent, they said, was too out-there — and really corrosive to boot. The yields were very low, and the investment was high. Luckily, researchers on the threshold of a discovery aren't easily dissuaded, so they ignored the engineers.

Folks around the lab felt that a concentration of polymer greater than 10 percent would be too thick. But one bright researcher tried a 20 percent mixture *at a high temperature*. To everyone's surprise, it worked — and allowed the materials to be spun at much higher concentrations, making the process economically feasible.

A second important discovery involved the way in which the fiber is quenched with water to cool it as it comes out of a spinneret. This alert researcher realized that if they added an air gap between the spinneret and the water, stress on the fiber caused the polymers coming out to align in the same direction. When the fibers cooled down, they froze with that same alignment, resulting in a much stronger fiber.

This was a horse of a different color: PPD-T now deserved some corporate attention. The product was unusual, the process used to produce it looked scaleable, and the dollars and cents made sense. The manufacturing engineers admitted the value, accepted the risks of building a plant using hot sulfuric acid as a spin solvent, and jumped on board.

Things heat up

Now they were up against it: They had to actually make this thing work in a manufacturing environment. This is the part that gives researchers and engineers nightmares.

Dozens of people from all walks of (scientific and engineering) life were gathered together in the great state of Virginia. Unbeknownst to them, the serious hurdles were just beginning.

First, how would they dispose of the sulfuric acid after the fiber-spinning process? After many late-night pizzas, they decided to convert it into calcium sulfate (commonly known as gypsum — the stuff used in wallboard and cement manufacture).

The second hurdle involved concerns about one of the chemicals used to polymerize the PPT-D. Though it had not been found to be toxic in previous studies, they took the time to run some further studies. One study with some hapless rats showed that the chemical could indeed be carcinogenic (cancer-causing) in animals. DuPont took immediate steps to change the handling methods of the chemical so that no one working with it — or the community at large, or the customers — would be put at risk.

DuPont also started a search for an alternative chemical with low toxicity that could produce the same properties in the fiber. They found one in relatively short order, but it caused another problem.

Turns out that when you use the alternative chemical with lower toxicity, it produces not only the correct long polymer chains, but also some short polymer chains that have to be weeded out. The engineers had to get together over coffee and donuts to modify the process to get rid of these tiny polymers.

To market, to market . . .

Anybody who has ever brought a product to market knows that R&D and engineering are easy compared to figuring out whether anybody will actually *buy* the thing (a factor called *market potential*). The final step in the process of bringing Kevlar to light was building a full-scale plant — to the tune of $400 million. As the building progressed, people were desperately seeking practical applications for the fiber.

At the time, *Fortune* magazine called Kevlar "a miracle in search of a market." DuPont entered into lots of partnerships with potential customers to see what people actually needed from fabrics.

The first area of commercialization focus involved steel-wire ropes used underwater with pulleys in offshore oil drilling. Kevlar has a strength 20 times higher than steel in sea water. Everybody thought Kevlar ropes should be lighter and easier to handle and have a longer lifetime than steel. Surprise! Kevlar rope lifetimes ended up being only about 5 to 10 percent more than steel rope — not the major selling point everyone had expected!

People scurried to find a way to improve the rope. One thing they tried was to make the ropes out of three different diameters of fiber. This compacted the structure and spread out the load; the resulting rope lasted five times as long.

Another change involved lubrication of the strands. DuPont surrounded each cable in a braid impregnated with fluorocarbons. This did a lot to reduce friction, heat, and abrasion. The lifetime of the cable now got better by a factor of six.

The final challenge was to optimize the angle at which they twisted the fibers to form the rope. This provided additional improvements in the life of the rope — and led to the use of Kevlar for many more types of rope applications. And that was only the beginning.

Kevlar's birth gives a good example of the kinds of problems anybody working with new science and commercial realities has to face. But to give those budding scientists among you hope, after all the hardships, Stephanie Kwolek was inducted into the U.S. Inventor's Hall of Fame in 1995.

Jumping Over the Hurdles

Though nanotechnology holds great promise in many areas, fulfilling that promise could come at a price. Companies looking to bring nanotechnology-based products to market will face various hurdles.

The availability or cost of equipment used to monitor the quality of the product could be a problem. The smaller the product being worked on, the more sophisticated and expensive the equipment needed to monitor quality control. For example, you might need a pricey electron microscope to measure size variations in a product.

The equipment needed for producing nanomaterials may not be readily available — or could cost a bundle. Equipment that can perform operations small enough to make some products may not even exist yet, which could delay projects. (Details, details.)

We talk in more detail about microscopes used to monitor nanomaterials and nanomanipulators in Chapter 3 — and take a look at equipment used in the manufacture of computer chips in Chapter 6.

Then there are the regulatory hurdles: Evaluations and government approvals are needed to bring products such as a new drug to market. All that hoop-jumping could take a long — and we mean *long* — time.

The scarceness or cost of nano-scale raw materials (such as buckyballs) could limit what products are manufactured and when. Take a look at Chapter 4 to explore some of the nanomaterials that are used (or are expected to be used) in products — and the ongoing efforts to improve their availability.

We'll have to develop procedures to mass-produce products developed in labs. The factors affecting the development of the manufacturing process might include

- ✔ **The safety factor:** Making sure the chemicals used are safe in a manufacturing environment and meet all government regulations.

- ✔ **The awful-glop factor:** Having an environmentally safe way to dispose of or recycle used chemicals.

- ✔ **The heavy-duty factor:** Producing production and monitoring equipment able to withstand the rigors of full-time use with reasonable maintenance requirements. There can be a significant difference between the performance of a single piece of equipment that is pampered in the lab by doting scientists and a bank of the same machines run by operators on a production line.

- ✔ **The uniformity factor:** Developing a production process that minimizes variation during mass production so that enough of the products produced meet specifications. This goes hand in hand with the need to design the product so it functions the way it's supposed to, even with a typical level of variation found in the production process.

- ✔ **The quality-control factor:** Finding a way to test or screen out unacceptable products.

- ✔ **The expertise factor:** Locating enough human resources (for example, suitably equipped engineers and scientists) to troubleshoot the process when, inevitably, issues show up in mass production that never appeared in the lab. (This is in accordance with the rule that a problem with your car never happens when you show it to the mechanic.)

Looking at Ethics and Society

Not all challenges with nanotechnology are technical; sometimes the human challenges are at least as tough. For example, the development of nanotechnology entails two major ethical issues that we'll all have to grapple with at some point. The first is the possibility that nanomaterials could cause harm, either to the environment or to human beings. The second relates to the way in which nano-benefits will be divided among the nations of the world.

Possible harm from nanomaterials

We've seen the scenario in a slew of science fiction movies — whenever a scientist discovers something new, some hideous green monster forms out of the goop and attacks its creator. Nanotechnology may not suffer the same fate as Baron von Frankenstein, but it is playing around with some things we've never played with before, and that concerns some people.

Over the last ten years or so, work with nanomaterials has moved along smartly — and by now we're using them in a whole bunch of products such as semiconductor chips and drugs. Tests for toxicity that might result from using these materials have been much slower to appear. Still, some experiments have shown there is cause for concern — and several groups are expressing concern because hundreds of products using nanomaterials are already on the market and more are on the way. Nanotechnology is likely to become a trillion-dollar industry in less than ten years. With that kind of explosive growth, some kind of watchdogging is indicated.

In the wake of cautions about (for example) genetically modified crops, governments and corporations are paying closer attention to how nanomaterials affect the environment, humans, and animals. The U.K. Royal Society, that venerable scientific institution, has commissioned a study of the potential risks and benefits of nanomaterials. The U.S. Environmental Protection Agency, not to be outdone, has also funded studies into potential health and environmental effects of nanomaterials, as well as studies to evaluate the use of nanomaterials to *clean up* the environment. For more about these studies, visit the EPA's Nanotechnology Web site (`http://es.epa.gov/ncer/nano/`).

Here are some findings about nano that are causing a stir:

- ✔ DuPont has determined that Single-Walled Nanotubes (SWNTs) can cause lesions in the lungs of rats. In one study, 15 percent died from suffocation when clumps of nanotubes lodged in their lungs. The study indicates that size matters; nanoparticles are generally more toxic when inhaled than are larger particles of the same material.

- ✔ Because of the small size of nano materials, people have looked at the possibility that they could interact with living cells in ways we can't anticipate. Researchers at Southern Methodist University in Dallas have found that buckyballs can, in fact, disrupt the membranes of fish brain cells. Trying to establish whether the same thing could happen to people, a team at Rice University exposed human liver and skin cells (grown in the lab) to solutions containing buckyballs. They found that a diluted solution (20 parts per billion) was capable of killing half the cells. At this point, neither experiment has been truly conclusive, which means more research needs to be done.

The good news is that researchers have found that by bonding simple chemicals to carbon spheres, they could reduce toxicity. This finding could be helpful in enabling nano participation in drug delivery. However, ultraviolet light from the sun might reduce the effect, rendering the spheres toxic again.

Because of such findings, the Action Group on Erosion, Technology, and Concentration (ETC Group) is actively lobbying for a nanotechnology research moratorium — a halt to all research until all hazards are investigated. Rice University's Center for Biological and Environmental Nanotechnology has taken a different approach, launching the International Council on Nanotechnology (ICON) to encourage collaboration among academic, industry, regulatory, and nongovernmental interest groups to assess, communicate, and reduce potential environmental and health risks associated with nanotechnology. ICON is working in four areas, including social-science research and public communication about nanotechnology. Their members include, among others, DuPont, Procter and Gamble, L'Oreal, and Intel. Interestingly, when the group was formed, three environmental groups refused to become members, fearing corporate domination of the process.

A new field, nanoecology, is emerging to study the interface between molecular technologies and our natural environment

Encountering a Nano Divide?

No technology exists in an economic vacuum; nano is no exception. Its development requires high levels of investment and an already-advanced technology. What happens to countries that don't have those? Well, the development of HIV/AIDS drugs may offer a sobering example: When the rich countries of the world developed them, the poorer countries — whose need became even greater — couldn't possibly afford to buy them. Many people are worried about a similar divide occurring when the wealthier countries that are pioneering nanotechnology research file all the patents and reap all the rewards. Countries with less-educated workforces won't be able to compete in the nanotechnology-related future. Benefits in medicine and other areas may "follow the money" and not be shared equally.

Who gets to benefit?

The issue is at least under discussion. The Rockefeller Foundation, for example, co-sponsored a conference entitled Global Dialogue on Nanotechnology and the Poor. Leaders who attended expressed concern about access to revolutionary technological discoveries, and many spoke of the need to ensure that such discoveries benefit the poor as well as the rich.

Parallels to pharmaceuticals?

When do intellectual property squabbles give way to the demand to humanely meet people's needs? The pharmaceutical industry is currently battling this very question. It takes U.S. companies at least 12 years to get a drug developed and approved, but by year 20, the patent expires and generic drug makers cut into the original drug maker's profits. These profits are needed to recover costs of research, development, and testing, and to do further research for drugs to fight other diseases. On the other hand, the generic drug makers lower the price considerably, making the drug available to a wider audience. The early players in nanotechnology will unquestionably face some of the same tricky dilemmas.

The debate often gathers around the expectation that benefits of nanotechnology could directly address the woes of third-world countries — these, for example:

- **Potential uses of nanotechnology to detect AIDS and cancer:** Promising — but who will share in that benefit? Similar tests for TB and malaria, both much more prevalent in developing countries, could end up being unavailable where they are needed most.

- **TB diagnostic kits using nanotechnology:** These could reduce the time and cost of tests and save lives. The size of these devices could make them ideal for carrying into remote regions.

- **Vaccinations for children:** Advanced nano-medicine could save many lives as nanoparticles deliver medicines more effectively.

- **Nano-repair of damaged bones:** Potentially a boon in developing countries where traffic accidents have been called an "unseen epidemic."

- **Enzyme biosensors:** These could be useful in monitoring toxicity in soil and crops. This could improve agricultural quality control.

- **Environmental cleanup:** Nanomagnets might be used to suck up oil in large-scale oil spills.

- **Lower-cost energy:** The use of nano-based solar cells could power rural areas without having to build an extensive power grid.

- **Improving the water supply:** Nanoparticles could help to filter and clean polluted water.

Making the numbers work

In addition to the technical side, economic concerns crop up: Industrialized nations will certainly be the first to benefit, but as mass production takes hold, technology could trickle down to developing nations. Still, many people worry about who will control the means of production of many of these beneficial products. Who will be in charge — corporations or specific governments with more military power? Will developing countries have the resources to set up, monitor, and enforce safety regulations related to nanotechnology? Will richer countries invest their dollars in (for example) nano-based stain-resistant jeans, age-defying nano-cosmetics, and other "frivolous" products, ignoring the research into potentially life-saving materials that third-world countries are literally dying for?

There is also concern that advances in nanotechnology will undermine certain industries that support the economies of developing countries, such as the extraction of natural resources. For this reason, many are calling for rich nations to devote a certain amount of the investment dollars that go to overseas countries to nanotechnology research in developing countries.

Nano and the economy

The government committee that had the privilege of wading into the controversy associated with putting genetically engineered foods on the supermarket shelves of England is facing another heated public debate about nano-foods. The Action Group on Erosion, Technology, and Concentration (the same ETC Group we mention earlier in this chapter) is concerned not only that the potential harm of nano-foods and pesticides is unknown, but also that nanotechnology will have an adverse impact on farmers in poor countries.

Others worry about shifts in the supply and value chains in the world economy. Small farmers in developing countries will be among the first affected. For example, products such as rubber and cotton (produced primarily in poor countries) could be largely replaced by nano-based materials.

Whatever the outcome of the debate, there are real ethical and social questions that a revolutionary field such as nanotechnology poses that will have to be addressed on a global level.

Chapter 3

Gathering the Tools of the Trade

. . . .

In This Chapter

▶ Boning up on chemistry and physics basics

▶ Analyzing things with spectroscopy

▶ Getting to the atomic level with microscopy

▶ Manipulating nano-size objects

. . . .

*I*t's only by visualizing the basic structure of matter that you can get a good working sense of what's going on in the nano-world described in the rest of this book. So get ready for a field trip to the nano construction site for a look at how molecules are formed from atoms, how light interacts with matter, and what tools scientists use to study and manipulate things about the size of (naturally) a nanometer.

Consider this the chapter where you get some up-to-date science basics under your belt. If you're a whiz in chemistry and physics, skip ahead to the section "Analyzing the Composition of Objects with Spectroscopy," but for those of you whose science is a tad rusty, read on.

That Bit of Chemistry and Physics You Just Have to Know

Nanotechnology presents you with a world that's normally outside everyday practical experience. Before you take your first (ahem) tiny steps into the world of nano-size objects, it's a good idea to dig out your old physics and chemistry notes from high school and/or college. If you don't have those notes handy (the dog ate them, or you steered clear of chemistry and physics in school), don't worry: We provide the basics right here. First we cover the building blocks of matter, atoms, and molecules, and explain how they stick together. Then we offer enlightening remarks on light — the dominant way most of us perceive things around us — including how researchers can see tiny nanoparticles by using super-duper microscopes (that's a technical term).

Molecular building blocks

One basic that even Einstein started with is that our physical universe is made up of matter and energy. At its root, chemistry is primarily concerned with describing the physical matter in that universe. Chemistry asks questions like these:

- ✔ What is matter composed of?
- ✔ How will a substance behave under certain conditions?
- ✔ What combinations of substances and conditions allow us to make other substances?

Physics, on the other hand, has its own set of questions to deal with. Physicists spend their time studying the forces that range from those holding each individual atom together to forces that hold the entire universe together by binding matter and energy.

A basic grasp of both physics and chemistry is essential equipment if you really want to understand our universe on a tiny scale. To make yourself at home in this nano-realm, for example, you need to understand what matter is composed of.

Start with a basic building block of nature — atoms.

Looking at the atom from the inside out

To get the hang of how nature uses atoms as building blocks requires a basic understanding of how they're structured. Here's the anatomy of an atom:

- ✔ Atoms are composed of even smaller particles; the three main types of these *subatomic* particles are protons, neutrons, and electrons.
- ✔ Protons and neutrons cluster together in the center of the atom — its *nucleus.* Protons are positively charged; neutrons are neutral.
- ✔ Orbiting around the nucleus — in much the same way the moon orbits around the earth — are smaller, negatively charged particles known as *electrons.*

Putting these pieces together, you get an image like that shown in Figure 3-1 — the classic symbol of all things "atomic." Note that the size of the nucleus in relation to the atom is seriously exaggerated in this figure: If an atom was the size of the earth, the nucleus would only be the size of a bowling ball!

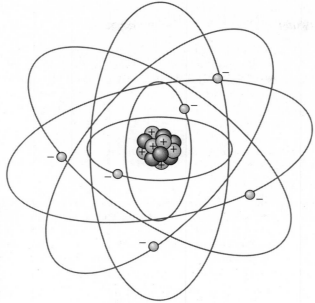

Figure 3-1:
A diagram
of the
structure of
an atom.

May the force(s) be with you

An attracting force known as the *strong nuclear force* holds protons and neutrons together in the nucleus of an atom. Despite its name, the strong nuclear force has nothing to do with The Force in *Star Wars* — in fact, it plays only a minor role in determining the actual structure of an atom. This rather muscle-bound force has only a limited range — it's very strong over very short distances (such as 10^{-15} m), but it disappears quickly at a range much beyond that. (It's nearly nonexistent outside the nucleus.)

From the atom's point of view, the more important force — the one that really gives the atom its form — is the *electromagnetic* or *electrostatic force* that acts between charged particles — you know, like those protons and electrons we just told you about. As you might expect, this force is greater at close range. The nearer two particles are, the more they feel it — and it gets weaker as they get farther away from each other.

Forces, like people, can be either attractive or repulsive, depending on the situation. When you have two particles with opposite charges — the positive proton and the negative electron, for example — the electrostatic force is *attractive* and the particles are drawn toward each other (it's one realm where opposites *do* attract). If, on the other hand, two particles have similar charges (for example, two electrons), the force is *repulsive* and the particles tend to move away from each other.

You can visualize the effect of these forces by remembering that time you spent in elementary school messing around with magnets. You tried to get the north poles of two bar magnets to touch each other and you marveled at feeling an invisible repulsive force hell-bent on pushing them apart. (The coolest part was when you released your grip on the two bar magnets and they skittered away from each other as if possessed.) When you turned one bar magnet around so the north pole of one bar magnet faced the south pole of the other, it was all you could do to keep the magnets from flying out of your hands in their rush to stick together.

Charged particles behave much like those magnets.

Holding atoms together

Practically speaking, how does an electrostatic force give an atom its shape and dimension? It's a kind of balancing act that works like this:

- The attractive force between electrons and protons pulls the electrons closer as they zip around the nucleus. This force prevents electrons from escaping from the atom.

- The repulsive force between the electrons themselves prevents them from hurtling toward the nucleus. That's because, as electrons move toward the nucleus, the volume the electrons occupy becomes smaller — and the distance between them shrinks until they remember that they can't stand each other.

- When electrons move closer together, the repulsive force gets stronger, forcing the electrons to move away from each other or stay as far away as they can get.

Ultimately the size of an atom is the result of two forces: the attraction of the nucleus for the electrons and the repulsive force between electrons.

Bonding atoms with electrons

Nature uses building blocks of many different sizes — not only atoms themselves but also stuck-together clumps of atoms (molecules). Just as something has to keep each atom in one piece, something else must hold the atoms together — just as mortar holds bricks together. In this case, that mortar is a relationship between atoms: chemical bonds.

Chemical bonding occurs when there aren't enough electrons to go around in the immediate vicinity, so the atoms get together and share electrons. This sharing allows each atom to reach a more stable state (which atoms like, being relatively lazy). An atom's outermost electrons are available for chemical bonding; they're the least strongly attached to their atom, so they're the only electrons that can be shared.

Each type of atom — hydrogen and helium, for example — has a different number of electrons available for chemical bonding and they are therefore more or less stable. That's why when you're around hydrogen you have to be careful not to burn things up, whereas you can safely use helium in a kid's birthday balloons. Because of the number of electrons it contains, helium is much more stable than hydrogen, so it disdains chemical reactions such as burning. (It's also great for situations when you just *have* to talk like a Munchkin.)

Moving on up to molecules

Some atoms have all the electrons they need, thank you very much, and prefer not to bond. They are the most chemically stable groups of atoms, the *noble gases* (such as neon and argon) — so stable they never feel the need to share electrons with other atoms.

Other types of atoms have to pool their electron resources to get greater stability, so they join together to form larger structures — *molecules.* For example, an oxygen molecule — which, unless you're an alien life form, you're breathing right this second — is just two oxygen atoms joined by a chemical bond; sharing electrons makes each oxygen atom more stable. Methane (which you want to avoid breathing at all costs) is one carbon atom joined with four hydrogen atoms by a chemical bond — sharing electrons makes all the atoms involved (both carbon and hydrogen) more stable.

When atoms join together via chemical bonds, you get the most stable molecules when each atom has as many electrons whizzing around it (counting the shared ones), as you'd find in a noble gas. In effect, every atom aspires to be as "noble" as possible but has to share. Result: molecules.

Molecules can range in size and complexity from a few atoms to many thousands of atoms. For example, a simple polymer (which you'll hear more about in subsequent chapters) is a long chain of several thousand methane molecules joined by chemical bonds between the carbon atoms, with two hydrogen atoms (rather than methane's standard four) per carbon atom.

Put enough molecules together, you get materials

Eventually, collections of molecules bond together to become things we use every day (like plastic and chocolate). But what determines these larger structures of molecules? Well, for openers, the same force that keeps an electron in an atom also sticks molecules together — the electrostatic force.

The strength of that *intermolecular* bond (which also qualifies as an attraction or a force) depends on how the electrons in the molecules are arranged. As the molecules get together, the intermolecular bond is a major factor in determining what kind of material is formed. Strong intermolecular bonds might give you a liquid at room temperature — weak intermolecular bonds might give you gas. (So to speak.)

Impurities can be useful, Mr. Bond

Consider the humble plastic-wrapped electrical wire: In the everyday world, it's common knowledge that the metal wire acts as a conductor and the plastic coating acts as an insulator. On the nano level, what's going on is a difference in methods of chemical bonding. The *metallic bonding* between the metal atoms allows the electrons shared between atoms to move from atom to atom — when these *delocalized electrons* are present, electricity flows. On the other hand, the *covalent bonding* between the plastic atoms forbids the shared electrons to move around. No delocalized electrons allowed; the plastic stops electricity in its tracks, keeping the zap in the wire where it belongs.

This relationship gets hazy when you tamper with the chemical bonds by adding impurities. For example, materials that are *semiconductors* (such as silicon) have electrical properties somewhere between conductors and insulators —

and can act either way, depending on the situation. When silicon is pure, it acts like an insulator; when it's not pure, it conducts different amounts of electricity, depending on how much of what stuff you add to it (a process called *doping*). Introducing different impurities creates different types of semiconductors. Down there in the nano realm, a negotiation among chemical bonds determines how freely the electrons can move.

When specific impurities — say, arsenic and boron — are added to semiconductors, they get silicon to act as a conductor, each in a different way: Adding arsenic (which contributes those footloose delocalized electrons) creates an "n-type" semiconductor; adding boron instead creates a "p-type." P-type semiconductors have no delocalized electrons; instead, they have "holes" — places where electrons should be, but aren't. Then current can flow because the electrons move from one hole to another.

Need a few classic illustrations of how intermolecular bonds affect a material's properties? (Doesn't everybody?) Try these on for size . . .

Somebody get a mop

In water molecules — each made up of one oxygen atom and two hydrogen atoms — the negatively charged electrons huddle near the oxygen atom, away from the hydrogen atoms. That gives the oxygen side of the molecule a negative charge and the side with the hydrogen atoms a positive charge. The result is a *polar molecule* (one side has a stronger positive charge and the other side has a stronger negative charge). The intermolecular bond between water molecules is strong enough to make a liquid at room temperature; it happens because the hydrogen atoms in one water molecule are drawn to the oxygen atom in another water molecule. In the nano-realm, it's all about electrostatic attraction between molecules; in the larger everyday world, things are just wet.

Oxygen — it's a gas!

Oxygen is breathable at room temperature because its intermolecular bond is feeble. Each oxygen molecule contains two oxygen atoms, with the electrons evenly distributed between them — symmetrically, in a *nonpolar molecule.* (Not quite a noble gas, but it does put on some airs.) If you distribute the positive and negative particles evenly, you get a puny electrostatic attraction — much weaker than for water — so not much intermolecular bond. No wonder the molecules bounce around the room.

The intermolecular bond between nonpolar molecules (the sort you find in oxygen) is called the *van der Waals force.* It's puny — totally dependent on the constant movement of electrons within molecules — which causes electrostatic attraction only part of the time. When the electrons move again, the attraction disappears. So, even though the van der Waals force is the same type of attractive force that occurs between polar molecules, it's only a part-time force — so the total attraction is much lower.

Beyond bonding, there's molecular structure

The type of bonding that goes on between molecules causes materials to have certain properties (such as conductivity). In addition, the way atoms and molecules fit together can affect a material's properties.

For example, graphite, which is used in lubricants and in pencil leads, is essentially sheets of carbon atoms bonded together into one huge molecule, with hydrogen atoms hanging on only at the edges. Because the only bonding between sheets is the van der Waals' force, the sheets slide easily over each other. Drag the graphite across paper, and it leaves a trail of itself on the page.

It just so happens that diamonds also contain carbon atoms — in fact, that's all a diamond contains — stacked neatly in a three-dimensional array or *lattice.* This three-dimensional structure has bonds from each atom to all the atoms immediately surrounding it — which makes the entire diamond one molecule with a 3-D structure. (Yep, one month's salary buys you one molecule — okay, a *gigantic* one — for your engagement ring.) One other result is some serious stability, which gives each diamond its strength. Figure 3-2 shows the structures of both diamonds and graphite.

The everyday world seems pretty solid to us, but deep in the nano-realm, matter is formed out of dynamic relationships and compromises — and now we can affect the outcome directly. You can see why nanotechnology researchers get excited about the ability to manipulate things at the atomic and molecular level. For example, you can customize substances. Imagine a future nanotechnology that starts out with just carbon atoms and ends up with superstrong, diamondlike, lightweight structural members for airplanes or spacecraft. It's closer than you might think.

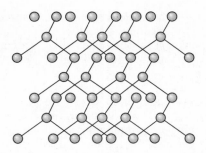

Diamond Structure

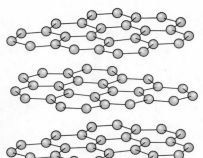

Figure 3-2:
Structures
of diamond
and
graphite.

Graphite Sheets

Turning on the light

Whenever you look at anything, from this book in your hands to a mountain range in the distance, you are essentially seeing light. Some things in nature produce light — such as the sun and lightning bugs — but only when (respectively) you're on the dayside of the planet or the bugs are looking for mates. After a few thousand years, people got pretty sick of sitting in the dark half the time, so they set about inventing their own devices to produce light — fire, light bulbs, glow-in-the-dark decals, the usual stuff that works pretty well in the everyday world.

The better we get at creating light, the more theories scientific types come up with to explain how light is generated, how it travels, and how it behaves. A quick rundown of the current best and brightest notions will shed light (sorry) on the last two sections of this chapter — the ones that deal with spectroscopy and microscopy, two of the most important tools used in nanotechnology research.

Thinking about light

For a long time, people thought of light as being made up of itsy-bitsy parti-cles, too small for anybody to see. Isaac Newton — a pretty smart guy, if truth be told — was one of the champions of *particle theory* (which stated that light is essentially a stream of such particles).

Some physicists disagreed with the particle theory, maintaining that light acted like (and so was composed of) waves. In the 1800s, in fact, some researchers showed conclusively that light had properties that could *only* be explained if it were made up of waves. This *wave theory* stated that light had properties similar to a wavelike electric field traveling with a wavelike mag-netic field. In light of this theory (sorry), light took its place in a spectrum of *electromagnetic radiation* that would eventually include heat waves, X-rays, and radio waves.

Along came Albert Einstein about a hundred years ago (in the early 1900s) with a revolutionary bright idea about the *photoelectric effect,* the tendency of some metals to give off electrons when struck with ultraviolet light. This effect is evidence that you can transfer energy into something by shining light on it.

At the time, wave theory couldn't account for all the details of this effect. Wave theory suggested that the photoelectric effect should cut off when the light you are shining on the metal falls *below* a certain intensity. (*Intensity* here is simply the amount of light present.) Instead, it was determined that decreasing frequency of the light (a measurement related to the amount of energy in a light wave) caused the cutoff, rather than intensity.

According to Albert, you can think of light as being made of little energy packets called *photons.* Using photons (which act like particles), you can explain how the photoelectric effect works. Light comes in different frequen-cies. Each photon contains a certain amount of energy related to the frequency of the light in it. An ultraviolet light is high frequency and high energy, while an infrared light is lower frequency and lower energy. If a photon has high enough energy (as with ultraviolet light), it can kick an electron loose from the metal and cause the current generated in the photoelectric effect. As light frequency goes down, the energy of the photons goes down and the current cuts off.

Today's physicists banter about both particle *and* wave theory when talking about light and generally believe that light can behave in both ways — as a particle *and* a wave — it depends on the situation. To describe light traveling from one place to another, we call on ideas from the wave model. When you talk about light interacting with matter on the atomic level, Albert's photons come into play — and into nano-research.

What exactly are light waves?

Consider, for a moment, a wave at the seashore. That wave occurs because energy is traveling through the water. The water on one side of an ocean doesn't move from one side to the other; instead, it's energy that moves across the ocean through the water. Light, in all its forms, is energy; it travels in waves.

Light waves don't even have to travel "through" anything; they are quite happy to travel through a vacuum. That's because light waves are made up of electric and magnetic fields — and those are energy, not matter.

Folks with nothing better to do measure the *wavelength* of light — the distance between one wave peak and the next (see Figure 3-3): The shorter the wavelength, the higher the energy; think about cooking things with microwaves — the *micro* means a really short wavelength. Way down at a tiny wavelength of 1 billionth of a meter, light waves are actually invisible, and referred to as *gamma waves* — a dangerous part of atomic radiation because of its high penetrating power. Up at the centimeter (or even meter) wavelength, you run into things like radio waves — still invisible but useful for broadcasting. What we can see with our eyes are wavelengths in the 400 to 700 billionths-of-a-meter range — just one small piece of the total spectrum of light.

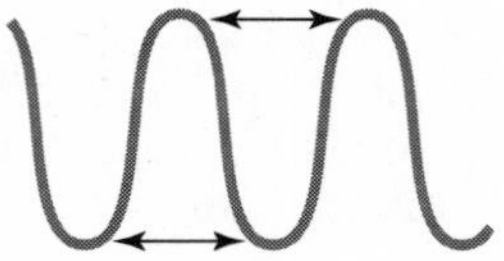

Figure 3-3:
Light waves
measured
in wave-
lengths.

Wavelength:
the distance between two successive
wave crests or troughs

Clocking light frequency

If you consider sound as a wave, then the vibration of the air is evidence of its passing by. If you measure how many waves go by in a second, you know (well, yeah) how frequently a wave goes by — so the name for this number is *frequency*. The unit of measurement that expresses frequency is *Hertz* (Hz), which means *cycles* (movement from one wave crest to the next) *per second*. Knowing how many crests are going by every second tells you how fast the light wave is moving.

Light you can see — called (well, yeah) *visible light* — is also called *color* because that's how we see it — in different colors at different frequencies. You see red at 430 trillion Hz and things look violet at 750 trillion Hz — and those extreme numbers are the two ends of the visible spectrum. When light goes above the top of the visible spectrum, it's *ultraviolet* (beyond violet); when it goes below the threshold of vision, it's *infrared* (below red) — either way, it's outta sight.

High-frequency light has a high level of energy; low-frequency light has a lower level of energy. To use the water analogy, the wavelength of light tells you how stretched-out or scrunched-up the individual waves are; the frequency tells you how often they hit. If the waves are all the same size, there's more energy delivered when more of them pound the shore (higher frequency).

But light doesn't just sit around; it also gets out and travels at various speeds. Light is generally a pretty fast little guy, traveling at a maximum speed of 186,000 miles per second (traffic cops be warned!) when racing through a vacuum — at its top speed, light is the fastest thing we know. When light travels through substances, however — even those we call transparent — it slows down. So light moving through water, a diamond, or even air moves more slowly than it does in the vacuum of space. The effect of different substances on light's speed can cause the beam of light waves to bend, which is called *refraction*. Light at different frequencies bends at different angles, which is what allows a prism to separate colors.

The whole *continuum* (that is, range) of light size, frequency, and energy is called the *electromagnetic spectrum* (illustrated in Figure 3-4). Visible light is only a very tiny part of that spectrum — about one-thousandth of 1 percent.

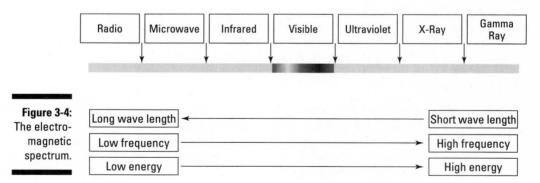

Figure 3-4: The electromagnetic spectrum.

Kicking out a photon

Watching light at play in the everyday world is pretty easy to get your mind around — but at the nano level, if you want to grapple with how light is generated or absorbed by atoms, you're better off thinking of light as a stream of photons. Every day your world is crisscrossed with zillions of photons coming from light sources such as lamps or the sun, reflected off of the things around you. Eyes absorb photons bouncing off the objects, triggering a nerve impulse to the brain, which registers the objects as "seen."

One way to produce a photon is to energize the electrons moving around an atomic nucleus — then something's got to give, and the atom gives off a photon to get back to its former peaceful condition. Now, normally the electrons that move around the nucleus are placed in the *orbital* (path around the nucleus) that requires the least amount of energy from them. But if you add energy to an atom, the electrons can move up to an orbital that takes more energy to maintain. But most matter is basically lazy; the electrons yearn for their cozy old orbitals, so the extra energy gets kicked out as a photon. Whenever an electron moves to a higher orbital and then falls back to its natural orbital, the excess energy is released *(emitted)* as light.

Sounds pretty busy, but each atom has to play by three rigid rules:

✔ Only photons with a certain specific amount of energy can be absorbed or emitted.

✔ Each electron can only enter certain orbitals.

✔ Each atom can only emit or absorb photons of certain frequencies.

In essence, it takes a certain amount of oomph to get an electron up to another (allowed) orbital from its hometown orbital. Each electron can absorb that much energy, no more. When the electron falls back to its familiar orbital, it emits exactly the same amount of energy, no more.

You're getting hotter . . .

So how do atoms get energized in the first place? In a word, heat. Heat up a piece of metal and it gets red. Get it even hotter and it turns white. Red is the lowest-energy visible light. When heat really gets the atoms going, all the colors in the visible spectrum appear at once — as white light. At the atomic level, all those excited atoms are emitting photons like mad.

An incandescent light bulb runs electricity through a filament — a wire designed to let its atoms heat up till they generate light. All kinds of man-made light sources do essentially the same thing in slightly different ways, from gas lanterns to fluorescent tubes. In the case of light sticks, a chemical reaction generates the needed heat. But at the bottom of all these methods is the need to energize those pesky atoms.

So what's the point in knowing how atoms act when they're excited? Well, for openers, you can figure out things at a distance — say, what's in a far-off planet's atmosphere — without having to go there. You can tell what kind of atom you're dealing with if you know what kind of light it gives off when it's energized — and that's as true for the nano realm as it is for the cosmic realm. Which leads us to a topic called spectroscopy, which allows you to determine what atoms are actually inside those nanomaterials.

Picking Apart Objects with Spectroscopy

The world of the very tiny is a constant bustle of activity. Things absorb, emit, bond, break up, vibrate, and travel, but they almost never stand around doing nothing (not even at temperatures near absolute zero). Of course, it's hard to catch them at it; nano-size things are too tiny to observe with the naked eye. Studying things that small requires special, deviously clever instruments that measure certain properties of matter — for example, *spectrometers* — tools that reveal the composition of things by measuring the light absorbed or emitted by atoms or molecules. Detecting and analyzing that data is the science of *spectroscopy,* and in this section we discuss three types of spectroscopy: infrared, Raman, and UltraViolet-Visible.

Infrared (IR) spectroscopy: Feel the heat

Vibration occurs in molecules because the chemical bond between atoms in the molecules works like a spring, as shown in Figure 3-5. The strength of that spring determines how much the atoms vibrate. Certain molecules absorb light at certain frequencies, which increases their vibration. The frequencies of light that can be absorbed are all in the infrared range and therefore the method of analyzing this vibration is called *infrared spectroscopy.*

Because atoms come in different shapes and sizes, and bonds have different strengths, each combination of atoms — and each type of bond — has a different *harmonic frequency* (a signature vibration that can be used to identify the composition of a material).

Here's how it works: Infrared light oscillates at the same frequency range as a vibrating molecule. When you shine infrared light on a vibrating molecule, the molecule absorbs only the frequencies that match its vibration. The energy absorbed from the infrared light exaggerates the molecule's movement. The part of the light that isn't absorbed can be measured by an infrared spectrometer, telling you which frequencies the object absorbed. That's a big clue to the object's composition, even if it's a really tiny object. Speaking of which . . .

Infrared spectroscopy just happens to be extremely handy for the study of single-walled nanotubes (SWNTs), which are discussed in Chapter 4. Understanding the bonds between SWNTs and materials like polymers is the key to building superstrong materials.

Figure 3-5:
A bond
between
two atoms is
like a spring
between
two balls.

You can use infrared spectroscopy to obtain information about some types of molecules but not others. If the molecule won't absorb frequencies in the infrared (IR) spectrum, you're out of luck; you have to use other types of analysis. IR spectroscopy can identify polar molecules (the type that act like magnets), but it can't identify nonpolar molecules. (For more about polar and nonpolar molecules, see "Put enough molecules together, you get materials," earlier in this chapter.)

Raman spectroscopy: Where's the energy?

Nope, Raman spectroscopy is not the inspection of Chinese noodles (and anyway that's spelled *ramen*). As with Infrared spectroscopy, the energy in photons used in Raman spectroscopy can be absorbed by the chemical bonds between atoms, exaggerating the molecule's movement. However, when the chemical bonds relax, going back down to a lower level of movement, or vibration, they release a photon. The level of vibration that the bond relaxes to is what determines the frequency of the newly released photon. For example, if the bond either relaxes back to its original state there is no valid data. If, on the other hand, the bond relaxes to a state that uses less energy for vibration than the original state, then the photon that is released is a higher frequency photon than the light used to excite it. If the bond relaxes to a state that uses more energy for vibration than its original state, the photon released is a lower frequency photon than the light used to excite it. Analyzing the change in frequency from the original photon not only tells you what kind of molecules are in a sample, but also tells you about the condition of the sample. That's because specific molecules absorb specific amounts of light energy — and the amount absorbed will change when they're under strain.

Essentially, you shine a laser at a sample and analyze the light it reflects. A laser is the best light source for this work because it produces only one frequency of light — which means you can ignore any scattered light that has the same frequency as the source. You're looking for light with a different frequency than the source.

Raman spectroscopy can identify nonpolar molecules; IR spectroscopy can't.

Of course, if you want to make a material stronger, you have to find out how it handles strain. In nanotechnology, you can probe the strains you put on a structure by using Raman spectroscopy to measure how the strain changes the vibration of the material. For example, if you're trying to develop composites in which nanotubes are bonded to another material such as a polymer, Raman spectroscopy can tell you how the strains on the nanotubes are distributed — must-have information if you're going to improve your composites.

UltraViolet-Visible spectroscopy: Who's there?

Molecules can also absorb ultraviolet light — and even visible light. So naturally somebody figured out how to use those two segments of the electromagnetic spectrum for analysis. The result — *UltraViolet-Visible (UV-Vis) spectroscopy* — uses either the UV or the visible part of the electromagnetic spectrum to put energy into a sample. Although the molecules in the sample could care less about the light, the *electrons* get all excited. The electrons in each type of atom can only absorb light of certain frequencies — as we discuss in the section "Kicking out a photon." The spectrometer measures that frequency of light that passes through the sample, and — voilà! — the frequencies that are missing reveal the identities of the atoms and molecules in the sample. This method is often used with molecules that IR spectroscopy can't identify; it's another window into the nano realm.

UV-Vis spectroscopy plays a role in the creation of *nanosensors* that can detect a material and identify its composition — by bonding with it (also called *capturing*), which changes the nanosensor's properties in specific ways that tell the tale. If you're developing a nanosensor over this next weekend or so (and who isn't?), you can use UV-Vis spectroscopy to verify the type of molecule that your particular nanosensor will capture.

Seeing Molecules with Microscopy

First, that microscope you used in fourth-grade science class to look at a super-big picture of a tiny blob of blood just won't do the job. Microscopy as used in nanotechnology goes way beyond that simple machine and involves some extremely sophisticated equipment. These microscopes allow you, through various methods, to view the surface features (referred to as *topography*) of atoms. Believe it or not.

Atomic force microscope (AFM)

If you're old enough to remember what a phonograph record was, you know that a crystal-tipped stylus ("needle") would move along in the grooves cut in a spinning vinyl platter, and when the motion vibrated the needle, the machine translated that vibration into sound. In a similar way, an *atomic force microscope (AFM)* scans the movement of a *really* tiny tip made of a ceramic or semiconductor material as it travels over the surface of a material, as shown in Figure 3-6. When that tip, positioned at the end of a *cantilever* (a solid beam), is attracted to or pushed away from the sample's surface, it deflects the cantilever beam — and a laser measures the deflection. The AFM then produces a visible profile of the little hills and valleys that make up the sample's surface.

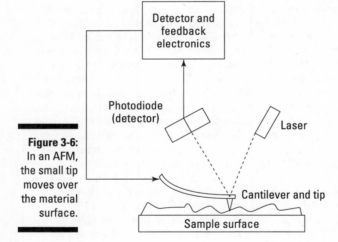

Figure 3-6:
In an AFM, the small tip moves over the material surface.

Figure 3-7 illustrates the surface typology shown by an AFM; the hills you see here represent atoms.

An AFM can do something you can't do with the electron microscopes we discuss in the next section: It can get images of samples in air and underneath liquids. That's because an electron microscope requires that the sample be in a vacuum — but an AFM does not. Also, an AFM produces a three-dimensional image; electron microscopes produce only two-dimensional images.

The fineness of the tip used in an AFM is an issue — the sharper the tip, the better the resolution. That's why scientists hope that the future developments bring the means to produce even sharper AFM tips. It's a natural for nanotechnology: Use of carbon nanotubes as tips for AFMs is in its infancy, but at least one vendor — Nanoscience Instruments, Inc. — already makes them.

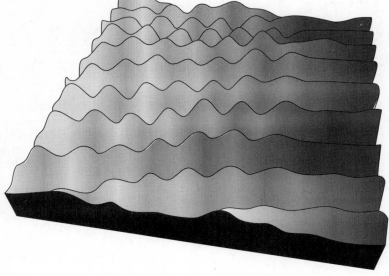

Figure 3-7:
An AFM
plot,
showing a
surface at
the atomic
level.

Using carbon nanotubes for AFM tips has the following benefits:

✔ Because the nanotube is a cylinder, rather than a pyramid, it can move more smoothly over surfaces. Thus the AFM tip can traverse hill-and-valley shapes without getting snagged or stopped by a too-narrow valley (which can be a problem for pyramid-shaped tips). Because a nanotube AFM tip is a cylinder, it's more likely to be able to reach the bottom of the valley.

✔ Because the nanotube is stronger and more flexible, it won't break when too much force is exerted on it (as some other tips will).

Techniques to make such tips are so new that they're not yet fully developed. Once they're perfected, however, nanotube tips are sure to make AFMs even more valuable tools for evaluating nanomaterials.

Scanning electron microscope (SEM)

You can use a *scanning electron microscope (SEM)* — which creates images of invisibly tiny things by bombarding them with a stream of electrons — to look at features on a scale as small as 10 nanometers (billionths of a meter). Figure 3-8 shows an SEM image of a nanotube hanging from one of the pyramid-shaped tips used in atomic force microscopes.

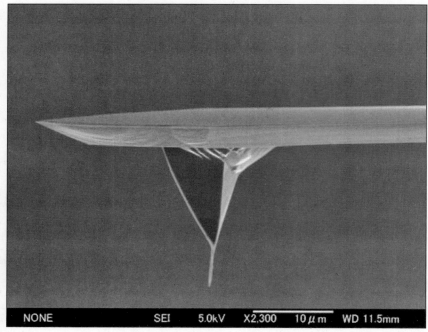

NONE   SEI   5.0kV   X2,300   10μm   WD 11.5mm

Figure 3-8: SEM image of a nanotube hanging off the tip of an AFM cantilever.

courtesy of Xintek, Inc.

An SEM shoots a beam of electrons at whatever you're examining, transferring energy to the spot that it hits. The electrons in the beam (called *primary electrons*) break off electrons in the specimen. These dislodged electrons (called *secondary electrons*) are then pulled onto a positively charged grid, where they're translated into a signal. Moving the beam around the sample generates a whole bunch of signals, after which the SEM can build an image of the surface of the sample for display on a computer monitor.

SEMs can ferret out quite a bit of information about the sample:

- **Topography:** surface features such as texture

- **Morphology:** shape, size, and arrangements of the particles that compose the object's surface

- **Composition:** elements that make up the sample (This can be determined by measuring the X-rays produced when the electron beam hits the sample.)

Transmission electron microscope (TEM)

Bouncing electrons off a sample is only one technique; you can also shoot electrons through the sample and watch what happens. That's the principle behind a *transmission electron microscope (TEM)*. In effect, it's a kind of nano-scale slide projector: Instead of shining a light through a photographic image (which allows certain parts of the light through), the TEM sends a beam of electrons through a sample. The electrons that get through then strike a phosphor screen, producing a projected image: Darker areas indicate that fewer electrons got through (hence that portion of the sample was denser); lighter areas are where more electrons got through (that's where the sample was less dense).

A TEM can achieve a resolution of approximately 0.2 nanometers, roughly the size of many atoms. Because most atoms have diameters of at least 0.2 nanometers, a TEM can produce images that show you just how the atoms are arranged in a material.

Although a TEM can achieve much greater resolution than an SEM, the instrument itself is costlier — and much more work is required to prepare a sample for a TEM than for an SEM. That's why computer-chip manufacturers use SEMs as everyday workhorses and only fire up the TEMs for more specialized measurements.

Typically, you'd use TEMs to analyze the morphology, *crystallographic structure* (arrangement of atoms in a crystal lattice), and composition of a sample.

The scanning tunneling microscope

Sounds like something out of Jules Verne, doesn't it? But if you want to get images of solid surfaces that conduct electricity, you can't beat the oddly named *scanning tunneling microscope (STM)*. It offers the best resolution of the bunch. This method uses an electric current (called a *tunneling current*) that begins to flow when a very sharp tip moves near to a conducting surface and hovers at about one nanometer away, as illustrated in Figure 3-9.

The tip (about the size of a single atom) sits on a piezoelectric tube. When you apply voltage to electrodes attached to this tube, you can make teensy adjustments to keep the tunneling current constant — which also keeps the tip at a constant distance from the sample while an area is scanned. The movement of the piezoelectric tube is recorded and displayed as an image of the sample surface.

Using a scanning tunneling microscope, you can see individual atoms on the surface of a sample — in 3-D. This technique is used to study things such as conductive materials and even DNA molecules.

What's a piezoelectric tube?

If you put electrodes on the opposite sides of some crystals — quartz or topaz, for example — and apply a voltage across the crystal, it will expand or contract. Any movement of the crystal in response to a voltage is called the *piezoelectric effect*. The piezoelectric tube used in the scanning tunneling microscope is simply a crystal that expands or contracts depending upon the voltage you apply to it.

An STM doesn't need to operate in a vacuum, which means you can use it to analyze samples in the air and in liquids. Often, however, a vacuum environment is used anyway to prevent samples from getting contaminated.

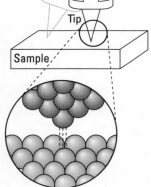

Figure 3-9: The tip used in an STM is sharp enough to narrow down to one atom at the point.

Magnetic Resonance Force Microscopy (MRFM)

MRFM is not only a mouthful to say (Magnetic Resonance Force Microscopy), it's a brand-spanking-new technique that takes advantage of the atomic-level vibration going on in just about everything. Still under development at the time of this writing (but worth watching), MRFM uses equipment similar to an AFM, with an important difference: The tip is made of magnetic material, and a special coil in the instrument applies a radio-frequency (RF) magnetic field.

Applying the RF field generated by the coil changes a quality called "spin" in the protons and electrons of the sample — flipping that quality back and forth. Each flip changes the magnetic field generated by the atoms in the sample — and the magnetic tip of the MRFM instrument moves in response. Result: the tip and the cantilever vibrate, which deflects a laser beam pointed at the sample. Using that deflection, the microscope detects the vibration and produces a picture of the atoms in the sample.

The microscope used in MRFM can vary the frequency of the coil's RF field. By mapping the location of atoms that respond to particular frequencies, you can (theoretically, anyway) find the precise location of specific types of atoms in a sample, which gives you (for starters) a fine-tuned sense of the sample's composition — at the atomic level.

Why are people so excited about MRFM? Because MRFM has a high resolution, can go beneath the surface of a sample, and (at least theoretically) can show atomic-scale 3-D images. Reason enough to get excited, we'd say — especially if you're doing nanotechnology research and development.

MRFM can be used in materials enhancement, physics, chemistry, biology, and medicine. Because of its high resolution and pinpoint accuracy, that may be just the beginning.

Moving the World with Nanomanipulators

Building sand castles as a kid was easy because you could scoop up millions of grains of sand in one hand and pack them all together. But imagine building a sand castle one grain of sand at a time.

That's kind of what nanotechnologists are up against. What they need to devise is a way to move around objects as tiny as a nanometer — lots and lots of them. Like the shovel and pail you used on the beach as a kid, so-called *nanomanipulators* are important tools used in the R&D of nanotechnology.

What's available today

One way to manipulate objects whose dimensions are measured in nanometers — such as carbon nanotubes — is to use atomic force microscopy (AFM). (See the previous section for more on AFMs.) Although an atomic force microscope is designed primarily to measure objects, it turns out that you can also use it to move tiny objects — such as those aforementioned carbon nanotubes.

Companies are now selling systems as add-ons to AFMs that increase their capability to manipulate objects. Some big players here are

- **3ʳᵈ Tech, Inc.**, developer of the nanoManipulator system, which gives your garden-variety AFM a virtual-reality interface. Equipped with one of these jobs, when the AFM tip skims over the surface of a sample, the tip/sample interaction is monitored. This yields information about properties such as friction and topography. Then a computer creates a graphic image of the surface. Using an attachment called a force-feedback pen (attached to the scanning tip), you can touch the sample surface and manipulate objects on it. Using this device, researchers have been able to manipulate carbon nanotubes, DNA, and even viruses. They can bend and rotate these objects, allowing researchers to find out more about the mechanical properties of said objects. The more you know about those properties, the better your chances of controlling the object.

- **MIT** has developed a nanomanipulator called the HexFlex. Well, no, it doesn't use witches' spells, but it *can* move samples a distance of 3nm (3 billionths of an inch) at a time — and that's pretty close to magic. The interesting thing about its design is that there are no joints, as there would be in a conventional manipulator arm. Instead, the main component is a flat six-sided piece of aluminum that is flexed by exerting force on one of its corners. MIT is developing another nanomanipulator called the MicroHexFlex. This one is only one millimeter wide. It's designed to reduce the loss of energy in fiber-optic systems by improving alignment between components.

What's down the road

Some people are never satisfied. Give 'em the capability to move material with the precision of a few nanometers or tens of nanometers, they complain it just isn't good enough. They have a vision of manipulating atoms and molecules one by one — never mind that this requires accuracy down to a *tenth* of a nanometer.

If you look at today's manufacturing methods from a nano perspective, they are as crude as the tools our caveman ancestors fashioned by chipping away at stone. When you mill a piece of metal, you fling molecules around here and there — at the nano level, there's a great lack of precision (sort of like using a nuke to dig the basement for a house).

In chemistry, things are rosier. Folks can build crystals that have all the atoms in the right places. We can specify the design of proteins or amino acids atom by atom. We can even copy DNA strands with incredible accuracy. However, our nano-abilities with manufacturing and electronics have a way to go.

When it all converges . . .

Down the road, molecular manufacturing is sure to evolve, allowing a precision in making components for computers or cars — molecule by molecule — that will rival the achievements of present-day chemists. Of course, to assemble such items in tiny pieces, we will need equally tiny robot devices. Things will begin at the nano level, and then be built up layer by layer. *Convergent technology* is the idea that you start small and by building hierarchically (level by level), you get bigger results. If you start with robots building nanometer-sized parts from individual atoms or molecules, and each group of robots builds parts double the size of those built in the previous stage, it should take about 30 stages to assemble meter-size parts.

Fat fingers, sticky fingers

Currently, nanomanipulators aren't capable of manipulating individual atoms and molecules, or of placing them with the precision required to assemble them into a manufactured product. Two of the challenges that scientists face are phenomena that Richard E. Smalley, 1996 Nobel Prize winner in Chemistry and currently teaching at Rice University, has identified. He calls these phenomena *fat fingers* and *sticky fingers*. Smalley explains that you can't make the fingers of a manipulator arm, which are made out of atoms, any smaller. At some point all those fingers simply don't fit in a nanometer-size reaction space. If you can't fit a certain number of fingers in the space, the manipulator can't control the chemistry.

The "fingers" of this manipulating nanobot would not only be too fat, they'd be too sticky, according to Smalley. The atoms that make up the manipulator hands could stick to the atom they're trying to move, which would mean that releasing an atom precisely could end up being a tricky business.

While some concepts have been proposed, nobody knows what these nanomanipulators will look like or what they will even be capable of, but everybody is looking forward to finding out!

Part II

Building a Better World with Nanomaterials

In this part...

*N*ope, nanotechnology isn't magic — it just sounds that way — though anything that can change the properties of materials at the molecular level does smack of alchemy. Stronger, thinner, more resilient, more conductive materials — you name it; people claim it as an anticipated benefit of nanotechnology.

In this part, we get down to the nitty—gritty of this brave new world. We discuss the building blocks of nanotechnology in Chapter 4 — including the carbon atom, carbon nanotube, and nanowires — and reveal how they're being used in products even today. In Chapter 5, we tackle the topic of composites — materials made out of nanomaterials that are evincing an astounding array of properties in all kinds of products.

Chapter 4

Nanomaterials Galore

Read the popular science journals out there and you'll see a lot of ink slung about how things made with nanomaterials are going to be stronger, more sensitive, lighter in weight, or have more load capacity than regular old materials. Delve into the issue a bit more, and you find out that actually getting simple carbon atoms to *do* all this for us is rather a complex little story.

To understand how nanomaterials will be used, you need a clear look at not only how they are formed but also their various configurations. In this chapter, we look at nano building blocks, and how they're currently being cajoled into enhancing all kinds of materials and products. Some of this work is still in the research phase; other work has graduated to the big bad world of real, existing consumer products and applications.

It All Starts with Carbon

Carbon atoms are all over the place. In fact, you can find them in millions of molecules. These molecules have a wide range of properties, meaning they pop up in every possible form — from gases such as propane to solids such as diamonds, the hardest material found in nature (and reportedly a girl's best friend).

"Bond, *carbon* bond . . ."

In covalent bonding, the atoms that bond share two electrons, regardless of whether they're shaken or stirred; this sharing of electrons is what holds the atoms together in a molecule. If the ability of each atom to attract all those negatively charged electrons (called *electronegativity*) is reasonably close (that is, if the difference in electronegativity is no more than 2), then they can form covalent bonds. Because the electronegativity of carbon atoms is 2.5 (roughly in the midrange), they can form strong, stable, covalent bonds with many other types of atoms with higher or lower values.

There are three significant reasons for the wide range of properties of materials containing carbon:

✔ **Carbon atoms can bond together with many types of atoms, using a process called *covalent bonding (we discuss the details of covalent bonding later in this section).*** When carbon atoms bond with different types of atoms, they form molecules with properties that vary according to the atoms they've bonded with.

✔ **Each carbon atom can form these covalent bonds with four other atoms at a time.** That's more bonds than most other atoms can form. Each nitrogen atom (for example) can form only three covalent bonds, each oxygen atom can form only two covalent bonds, and so on. This four-bond capability allows carbon atoms to bond to other carbon atoms to make chains of atoms — and to bond with other kinds of atoms at various points along such chains. This wide range of potential combinations of atoms in a molecule allows for a correspondingly wide range of potential properties.

✔ **There's no other element in the periodic table that bonds as strongly to itself and in as many ways as the carbon atom.** Carbon atoms can bond together in short chains, in which case they may have the properties of a gas. They may bond together as long chains, which might give you a solid, like a plastic. Or, they can bond together in 2- or 3-dimensional lattices, which can make for some very hard materials, such as a diamond.

With capabilities like these, carbon atoms are a natural for use in nanomaterials (as you discover in the following sections).

How Carbon-Based Things Relate to Nanotechnology

Buckyballs and carbon nanotubes are two types of molecules composed of carbon atoms that have many applications in nanotechnology. (Peruse the index of any book on nanotechnology — including this one — and you'll find umpteen references to buckyballs and nanotubes, a sure sign that they are the Big Kahunas of the nanotechnology field.) We give you the lowdown on both types of molecules later in this chapter. For the moment, however, we tell the tale of benzene and graphite (two natural carbon-based materials). A look at how their molecular structure influences their properties can help explain how the same thing happens with buckyballs and nanotubes.

Delocalizing with benzene

Electrons in some carbon-based molecules are organized in orbitals that allow the electrons to travel between atoms in the molecule. These traveling electrons are called *delocalized* electrons. Carbon-based materials that have delocalized electrons can conduct electric current. One example is benzene — a ring of six carbon atoms, each with one hydrogen atom attached, and plenty of delocalized electrons. (Figure 4-1 shows a benzene molecule in all its glory.)

Figure 4-1: A benzene ring.

Delocalized electrons

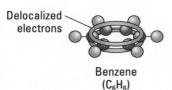

Benzene
(C_6H_6)

In benzene, each carbon atom uses one electron to bond with each of two neighboring carbon atoms and one hydrogen atom. This leaves one of the carbon atom's four electrons hanging around, unused. The atomic orbitals that contain these extra electrons overlap with orbitals on the adjacent carbon atoms, forming an orbital represented by the ring-shaped cloud above and below the ring of carbon atoms shown in Figure 4-1. This orbital allows the electrons to move freely around the ring of carbon atoms. (These are the delocalized electrons we told you about.)

Nanotechnologists have demonstrated that molecules such as benzene — because they contain these handy delocalized electrons — can be used to conduct electrical current in nano-scale electronic devices. (We discuss this process in more detail in Chapter 8.)

This bonding method was originally called *conjugated*. This term springs from a previous model of carbon bonding — which held that the extra electrons formed bonds between every other pair of carbon atoms. This situation would create alternating single bonds (where two electrons are shared) and double bonds (where four electrons are shared) between carbon atoms. Although scientists have replaced this model, some nanotechnologists still use the term *conjugated bond* to indicate the presence of extra electrons that can be used in a bonding process.

Letting things slide with graphite

Benzene gives you a simple example of how delocalized electrons work; based on that understanding, now you can see how a slightly more complex material, graphite (the tip of your #2 pencil), also conducts electricity.

A graphite molecule is, like benzene, a collection of carbon rings — but it's built differently: The same electrons that benzene uses to bond each carbon atom to a hydrogen atom are used instead to bond carbon atoms to other carbon atoms in adjacent carbon rings. What you get for a molecule is a sheet of interlocking hexagonal carbon rings, as shown in Figure 4-2. In this structure, each carbon atom bonds covalently to three other carbon atoms. The bonds are all carbon-to-carbon till you get to the edge of the graphite sheet, where the carbon atoms bond to hydrogen atoms (one each).

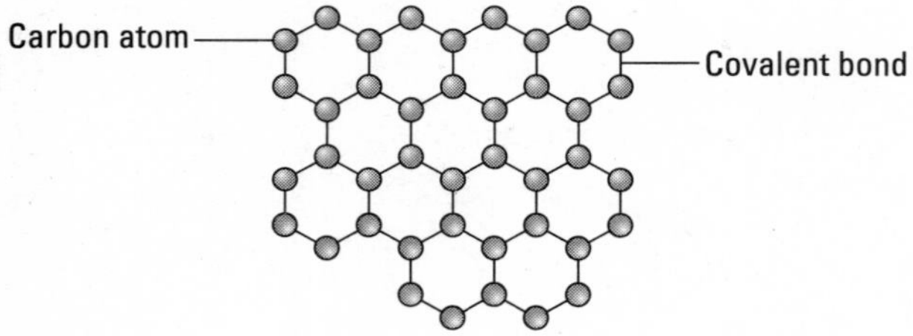

Figure 4-2:
The structure of carbon atoms connected by covalent bonds in a sheet of graphite.

Carbon atom —————— —————— Covalent bond

As with benzene, each carbon atom in graphite has an "extra" electron — one more than the number of atoms it's bonded to. The atomic orbitals for these electrons overlap to form a *molecular orbital* that allows delocalized electrons to move freely throughout an entire graphite sheet. That's why graphite conducts electricity.

A sheet of graphite is very strong because of the interlocking carbon-to-carbon covalent bonds. But wait a minute: The graphite in pencil lead doesn't *seem* all that strong (and it smudges all over you if you're not careful). That's because it's composed of many parallel sheets of graphite. Although each individual sheet of graphite is very strong, it's only one carbon atom thick — that's less than a nanometer. Because all the electrons have been put to use *within* the molecular sheet, the bonding *between* sheets of carbon is done with van der Waals' forces, which are very weak. The graphite sheets can slide across each other easily, which makes graphite useful as a lubricant or as a drawing tool — when you draw a picture of your house, sheets of graphite slide off onto the piece of paper.

Two properties of graphite sheets make them especially useful in nanotechnology: their strength and their ability to conduct electricity. Buckyballs and nanotubes are essentially graphite sheets shaped into spheres or tubes — and the next few sections provide a closer look.

Bouncing Buckyballs

A *buckyball* (short for *buckminsterfullerene*, a term we explain shortly) is a molecule containing 60 carbon atoms. Each carbon atom is bonded to three adjacent carbon atoms, just as in graphite. However the carbon atoms in a buckyball form a teensy-weensy sphere that's about 1 nanometer in diameter, as shown in Figure 4-3. Because one of the properties of carbon atoms is that they can bond to many other types of atoms, researchers can use them to create customized molecules, useful in various applications discussed throughout this book.

Figure 4-3:
Sixty carbon atoms in the shape of a sphere — a buckyball.

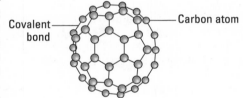

Covalent bond — — Carbon atom

Desperately seeking buckyballs

Buckyballs were discovered through an interesting collaboration of researchers from two universities. Richard Smalley at Rice University was studying semiconductor materials. He had a device that shined a laser at a solid sample, vaporized part of it, and analyzed the clusters of atoms that formed in the vapor.

Meanwhile, at the University of Sussex, Harry Kroto was attempting to reproduce a material found in deep space that generated specific molecular spectra from carbon atoms. Bob Curl, also from Rice University, was doing similar work. Curl ran into Kroto at a symposium and suggested he drop by Rice University because

Smalley's scientific instrumentation might be helpful in his work. Kroto dropped by Rice, and after seeing the work that Smalley was doing, he became interested in using that equipment to reproduce his carbon molecules.

Time spent working with high-end scientific equipment is always at a premium, so Kroto had to wait about a year until the equipment was available. In August of 1985, Smalley, Kroto, Curl, and some graduate students performed a series of experiments producing carbon molecules and clusters. They found that under certain conditions, most of the molecules generated contained 60 carbon atoms. Voilà: buckyballs.

While many of the atoms in buckyballs are connected together in hexagons (just as in graphite sheets), some of the atoms are connected together in pentagons. The pentagons allow the sheet of carbon atoms to curve into the shape of a sphere. Every buckyball surface contains 12 pentagons and 20 hexagons.

After much discussion and modeling, researchers determined that 60 carbon atoms form a single stable molecule only if they're arranged in 20 hexagons and 12 pentagons that are linked to form a sphere — as it happens, it's the same arrangement of hexagons and pentagons proposed by American architect and engineer Buckminster Fuller for his famed geodesic dome. No wonder these molecules got dubbed *buckminsterfullerenes* (that's *fullerenes* for short) in his honor. This type of spherical carbon molecule has been found in various other sizes. The fullerene family of molecules is often identified by the letter C followed by the number of carbon atoms, for example C60, C70, C80.

Creating buckyballs

Richard Smalley, one of the scientists credited with discovering buckyballs (see sidebar), produced the little wonders in a device by vaporizing carbon with a laser and allowing the carbon atoms to condense. However, this device could only produce a very small number of buckyballs — enough to show their existence but not enough to allow thorough study of their capabilities.

It took collaboration between researchers at the Max Planck Institute in Germany and the University of Arizona to figure out how to make larger quantities of buckyballs. Operating on a shoestring, they vaporized carbon by placing two carbon electrodes close together and generating an electric arc between them in a reaction chamber filled with a low pressure of helium or neon (neither of which will bond with carbon). This method generated much larger quantities of buckyballs than Smalley's device did, although the buckyballs had to be separated from carbon soot by using solvents such as benzene. The method produced enough buckyballs to analyze for their electrical and physical properties.

But what if you want to produce buckyballs by the truckload (as you'd need to for most commercial applications)? A little tech school in Massachusetts called MIT, working with a company called Nano-C, Inc., developed a method called *combustion synthesis,* which produces big enough quantities of buckyballs — at a low enough cost — for use in commercial applications. This method mixes a hydrocarbon with oxygen and burns the hydrocarbon at a low pressure. Nano-C states that its latest version of this process produces buckyball material pure enough (about 95 percent) to skip the step that separates the product from the carbon soot — and that reduces costs significantly.

Do you need buckyballs by the ton? The Frontier Carbon Corporation, a subsidiary of the Mitsubishi Corporation, has built a facility capable of producing 40 tons of buckyballs per year. They started production with an early version of the combustion-synthesis process developed by MIT and Nano-C, but have since switched to an undisclosed technique developed at Kyoto University.

Using buckyballs in the real world

At the time of this writing, most of the commercial applications of buckyballs are still in development. But just where does the most promise lie?

Buckyballs as antioxidants

The medical field is one place that buckyballs appear to have a promising future. C Sixty, Inc., is one of the companies developing medical applications for buckyballs. They are focusing on the ability of buckyballs to act as antioxidants, counteracting free radicals in the human body.

A *free radical* is a molecule or atom that has an unpaired electron — which makes it very reactive. An *antioxidant* is a molecule that can supply an electron and neutralize a free radical. The human body normally has a balance of free radicals and antioxidants; a certain level of free radicals is actually necessary to make your immune system work. However, the level of antioxidants found naturally in your body decreases as you get older. The resulting high level of free radicals roaming around your system could be the cause of certain diseases.

Buckyballs can act as antioxidants to neutralize free radicals. When a buckyball meets a free radical, the unpaired electron in the free radical pairs up with one of the buckyball's delocalized electrons, forming a covalent bond between the free radical and a carbon atom in the buckyball.

One problem with using buckyballs as an antioxidant is that antioxidants have to be soluble in water to be truly useful as medical applications. Buckyballs are *not* naturally soluble in water, and therefore not soluble in the bloodstream. (Remember that water makes up about 80 percent of blood.) To make buckyballs soluble, C Sixty has added a water-soluble molecule to them. This is done by covalently bonding an atom in the water-soluble molecule to one of the carbon atoms in the buckyball.

Bonding an atom or molecule to a buckyball to change the properties of the buckyball is called *functionalization*.

Merck, Inc., has obtained a licensing option on C Sixty's antioxidant, and is jumping through the various hoops of evaluation and qualification that the government requires before new drugs can be provided to the medical field.

They are going to all this trouble because studies have shown that buckyball-based antioxidants are several times more effective than antioxidants available today. Each buckyball-based antioxidant can counteract several free radicals because each buckyball has many carbon atoms for the free radicals to bond to. Antioxidant molecules currently in use can only counteract one free-radical molecule apiece; they have fewer places to which free radicals can attach themselves.

Improving medical imaging and drug delivery with buckyballs

Another potential use of buckyballs in medicine involves delivering elements for medical imaging. When you make a buckyball, you vaporize carbon. If you vaporize a metal along with the carbon, buckyballs conveniently form around the metal atoms. Fast-forward to the metal gadolinium, which is used for medical imaging, but with gadolinium there's a danger that some of the metal can remain in a body and cause damage. By containing these potentially harmful metal atoms in a buckyball, researchers hope to be able to flush the metal-enclosed-in-a-buckyball combination completely from the body after it has accomplished its medical imaging task, rather than leaving it there where it could be absorbed into tissue. (Chapter 10 details the use of buckyballs for medical imaging.)

Another use of buckyballs is to deliver drugs directly to infected regions of the body. It turns out that such regions have pH levels that differ from the pH of the healthy bits (pH measures the acidity of a solution). Researchers hope to functionalize a buckyball by bonding it to molecules that react to changes in pH, so the drug is only released at the infected area. In Chapter 11 we discuss this method in more detail.

One other reason buckyballs are useful for drug delivery is that they are small enough to move around inside the body quite easily.

Buckyballs at work everywhere

Here's a quick rundown of various efforts that are going on today to develop products using the capabilities of buckyballs. When these products will become available is anybody's guess at this point, but it depends in part on our ability to produce low-cost buckyballs.

- Dupont and Exxon are using buckyballs to develop stronger polymers. These companies are looking at two ways of doing this: by integrating the buckyballs in the polymer with chemical bonding, or by simply tossing buckyballs into the polymer and embedding them there.

- Additional buckyball-based antioxidant type drugs are being developed or tested. For example, anti-aging or anti-wrinkle creams are being developed by Mitsubishi, Taiwan University is testing a buckyball-based drug to fight arteriosclerosis, and C Sixty is working on both burn creams and an HIV drug.

✔ Sony is developing more efficient fuel cell membranes using buckyballs.

✔ Siemens has developed a buckyball-based light detector.

✔ Seagate is using buckyballs to develop diamond-hard coatings for computer disk drives.

Buckyballs Grow Up to Become Nanotubes

As far back as 1959, Roger Bacon had produced images of carbon nanotubes. In the 1980s, Howard Tennant applied for a patent for a method to produce them. In 1990, Richard Smalley postulated the concept that if buckyballs get big enough, they become carbon cylinders. But it wasn't until 1991 that Sumio Iijima, a researcher at NEC's Fundamental Research Lab, not only took photos of nanotubes, but also put two and two together to explain what nanotubes actually are — and put a name to them. He placed a sample of carbon soot containing buckyballs in an electron microscope to produce some photographs of buckyballs — which he in fact did — but some odd, needle-shaped structures caught his attention.

It turned out that these needle shapes were actually cylinders of carbon atoms that were formed at the same time that the buckyballs were formed. Like buckyballs, these cylinders (called *carbon nanotubes*) are each a lattice of carbon atoms — with each atom covalently bonded to three other carbon atoms. Carbon nanotubes are basically buckyballs, but the end never closes into a sphere when they are formed. Instead of forming the shape of a sphere, the lattice forms the shape of a cylinder, as illustrated in Figure 4-4.

Nanotubes come in a couple of varieties. They can either be single-walled carbon nanotubes (SWNT) or multiwalled carbon nanotubes (MWNT). As you might expect, an SWNT is just a single cylinder, whereas an MWNT consists of multiple concentric nanotube cylinders, as illustrated in Figure 4-5. We'll tell you about SWNTs rather than MWNTs, because most research is focused on developing uses for single-walled nanotubes.

The length and diameter of SWNTs varies, but a typical SWNT would be about 1 nanometer in diameter and a few hundred nanometers in length. The smallest diameter anybody has ever seen in SWNTs is about the same as the C60 buckyball, just under 1 nanometer.

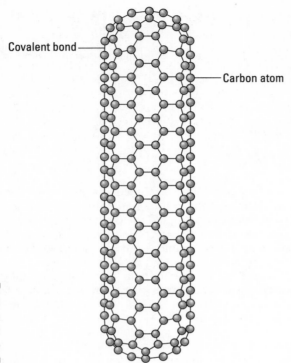

Covalent bond

Carbon atom

Figure 4-4:
Illustration
of a carbon
nanotube.

Figure 4-5:
Illustration
of a
multiple-
walled
carbon
nanotube.

Producing nanotubes from thin air

Researchers found that by adding just a few percentage points of vaporized nickel nanoparticles to the vaporized carbon (using either the arc-discharge or laser-vaporization method to produce the vapor), they could make as many nanotubes as buckyballs — or even more. Here's why: Carbon atoms dissolve in the metal nanoparticle. When the metal nanoparticle is filled to the brim with carbon atoms, carbon atoms start sweating onto the surface of the particle and bond together, growing a nanotube. When you anchor one end of the growing nanotube to the metal nanoparticle, it can't close into the sphere shape of a buckyball. This also allows the nanotubes to incorporate many more carbon atoms than a buckyball.

This method produces both single and multiple walled nanotubes intermingled with carbon soot. (If you want to know how to get nanotubes out of the marshmallows you roast over your campfire, flip over to Chapter 5, where we discuss methods of removing nanotubes from the carbon soot and untangling them.)

Once they discovered nanotubes, researchers set about trying to figure out ways to produce lots of them. There are three methods that various companies have developed to produce carbon nanotubes in bulk quantities and at a lower cost:

- The first method is called high-pressure carbon monoxide deposition, or HiPCO. This method involves a heated chamber through which carbon monoxide gas and small clusters of iron atoms flow. When carbon monoxide molecules land on the iron clusters, the iron acts as a catalyst and helps a carbon monoxide molecule break up into a carbon atom and an oxygen atom. The carbon atom bonds with other carbon atoms to start the nanotube lattice; the oxygen atom joins with another carbon monoxide molecule to form carbon dioxide gas, which then floats off into the air.

- The second method is called *chemical-vapor deposition,* or CVD. In this method, a hydrocarbon — say, methane gas (one carbon atom and four hydrogen atoms) — flows into a heated chamber containing a substrate coated with a catalyst, such as iron particles. The temperature in the chamber is high enough to break the bonds between the carbon atoms and the hydrogen atoms in the methane molecules — resulting in carbon atoms with no hydrogen atoms attached. Those carbon atoms attach to the catalyst particles, where they bond to other carbon atoms — forming a nanotube.

- A brand-new method uses a plasma process to produce nanotubes. Methane gas, used as the source of carbon, is passed through a plasma torch. Nobody's revealed the details of this process yet, such as what, if any, catalyst is used. One of the initial claims is that this process is 25 times more efficient at producing nanotubes than the other two methods.

Eying the structure of carbon nanotubes

A carbon nanotube is a cylinder of carbon atoms covalently bonded together, sort of like a sheet of graphite rolled into a cylinder. Some of these cylinders are closed at the ends and some are open. Each carbon atom is bonded to three other carbon atoms and forms a lattice in the shape of hexagons (six-sided rings of carbon atoms), except near the end. For nanotubes with closed ends, where the ends start to curve to form a cap, the lattice forms pentagons (five-sided rings of carbon atoms).

The lattice can be orientated differently, which makes for three different kinds of nanotubes. As shown in Figure 4-6, in *armchair* nanotubes, there is a line of hexagons parallel to the axis of the nanotube. In *zigzag* nanotubes, there's a line of carbon bonds down the center. *Chiral* nanotubes exhibit a twist or spiral (called *chirality*) around the nanotube. We discuss how the orientation of the lattice helps determine the electrical properties of the nanotube later in this chapter.

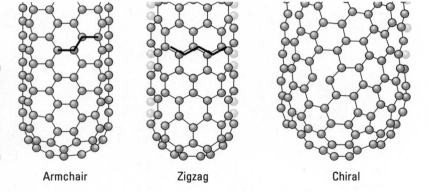

Figure 4-6:
Armchair, zigzag, and chiral nanotubes.

Armchair   Zigzag   Chiral

Scanning the properties of nanotubes

One property of nanotubes is that they're really, really strong. *Tensile strength* is a measure of the amount of force an object can withstand without tearing apart. The tensile strength of carbon nanotubes is approximately 100 times greater than that of steel of the same diameter.

There are two things that account for this strength. The first is the strength provided by the interlocking carbon-to-carbon covalent bonds. The second is the fact that each carbon nanotube is one large molecule. This means it doesn't have the weak spots found in other materials, such as the boundaries between the crystalline grains that form steel.

Nanotubes are strong but are also elastic. This means it takes a lot of force to bend a nanotube, but the little guy will spring right back to its original shape when you release it, just like a rubber band does. Researchers have used atomic force microscopes to physically push nanotubes around and observe their elastic properties. Evaluations with transmission electron microscopes — the kind of microscope sensitive enough to give you a peek at atomic shapes — show that the bonds in the atomic lattice don't break when you bend or compress a nanotube.

Young's modulus for carbon nanotubes, a measurement of how much force it takes to bend a material, is about 5 times higher than for steel, so if you were thinking of going out and bending a nanotube, think again. The fact is, there's not another element with a lattice structure in the whole periodic table that bonds to itself with as much strength as carbon atoms. And, since carbon nanotubes have such a perfect structure, they avoid the degradation of strength that you get with other materials.

In addition to being strong and elastic, carbon nanotubes are also lightweight, with a density about one quarter that of steel.

As if that weren't enough, carbon nanotubes also conduct heat and cold really well (they have a high *thermal conductivity*); some researchers predict a thermal conductivity more than 10 times that of silver — and if you've ever picked up a fork from a hot stove, you know silver and other metals are pretty darn good conductors of heat. While metals depend upon the movement of electrons to conduct heat, carbon nanotubes conduct heat by the vibration of the covalent bonds holding the carbon atoms together; the atoms themselves are wiggling around and transmitting the heat through the material. The stiffness of the carbon bond helps transmit this vibration throughout the nanotube, providing very good thermal conductivity.

A diamond, which is also a lattice of carbon atoms covalently bonded, uses the same method to conduct heat, so it's also an excellent thermal conductor, as well as a stunning piece of jewelry.

Carbon nanotubes are a little bit sticky, as well. The electron clouds on the surface of each nanotube provide a mild attractive force between the nanotubes. This attraction is called van der Waals' force (which we discuss in Chapter 3). This involves forces between nonpolar molecules (a molecule without a positive end and a negative end). A carbon nanotube just happens to be a nonpolar molecule.

Metallic or semiconducting?

The diameter of a carbon nanotube and the amount of twist in its lattice determines whether it's metallic or semiconducting. Electrons in carbon nanotubes can only be at certain energy levels, just like electrons in atoms. A nanotube is metallic if the energy level that allows delocalized electrons to flow between atoms throughout the nanotube (referred to as the *conduction band*) is right above the energy level used by electrons attached to atoms (the *valance band*). In a metallic nanotube, electrons can easily move to the conduction band. A nanotube is semiconducting if the energy level of the conduction band is high enough so that there is an energy gap between it and the valance band. In this case, additional energy, such as light, is needed for an electron to jump that gap to move to the conduction band. While there is no gap between the valance and conduction bands for armchair nanotubes (which makes them metallic), an energy gap does exist between the valance and conduction bands in about two thirds of zigzag and chiral nanotubes — which makes them semiconducting.

Obviously, it's important to be able to control what type of nanotube you are growing. Most current production processes for nanotubes create both metallic and semiconductor nanotubes. Researchers at Rice University have hit on a way to control this process. They take short lengths of nanotubes of the type they want and attach nanocatalyst particles (typically a metal such as nickel) to one end. These nanotubes are placed in the reaction chamber and act like seeds. New, long, nanotubes are grown from these seeds, kind of like nanotube cloning.

But not all nanotubes are exactly alike. Armchair nanotubes all have electrical properties like metals — but only about a third of all zigzag and chiral nanotubes have electrical properties like metal; the rest (roughly two thirds) have electrical properties like semiconductors. (For more about the difference, see the "Metallic or semiconducting?" sidebar in this chapter.) A metallic carbon nanotube conducts electricity when you connect different voltages to each end, just like a wire. Applying a negative voltage at one end and a positive voltage at the other end causes electrons to flow towards the positive voltage. To get electrons to flow through a semiconducting carbon nanotube, you also have to add some energy. (You could shine light on the nanotube, for example.)

Carbon nanotubes conduct electricity better than metals. When electrons travel through metal there is some resistance to their movement. This resistance happens when electrons bump into metal atoms. When an electron travels through a carbon nanotube, it's traveling under the rules of quantum mechanicals, and so it behaves like a wave traveling down a smooth channel with no atoms to bump into. This quantum movement of an electron within nanotubes is called *ballistic transport*.

Carbon atoms in nanotubes, like those in buckyballs, have the ability to covalently bond to other atoms or molecules creating a new molecule with customized properties. Bonding an atom or molecule to a nanotube to change its properties is called *functionalization*.

Putting nanotubes to good use

Researchers at Rice University are developing a way to form wires out of nanotubes. The process used to spin polymer fibers together to make Kevlar cables was used as a model to figure out how to spin nanotubes together into a wire. The wire is a bunch of metallic armchair nanotubes bundled together.

In a wire made of carbon nanotubes, electric current could zip through like a skater on ice. In addition, less energy would be wasted as heat. It would also weigh less than conventional wire, while being able to conduct huge doses of current. The impact of such wire on energy technologies could be big.

Researchers are also looking into using carbon nanotubes to detect chemical vapors. The way this would work would be that molecules that make up a chemical vapor would land on the nanotube and attach themselves to it by forming covalent bonds with the carbon atoms in it. This would change the electrical conductivity of the nanotube by decreasing or increasing the number of delocalized electrons available for conduction, which would then trigger an alarm of some kind.

Sensors using carbon nanotubes have been shown to detect chemical vapors with concentrations in the parts per billion (ppb). One problem is that many molecules can interact with carbon nanotubes in this way. In order to ensure that sensors are detecting the right chemical, researchers coat nanotubes with a polymer that allows certain molecules to reach the nanotube and blocks others. (We discuss nanotube-based chemical-vapor sensors in Chapter 8.)

What's in a name?

Electrons travel ballistically within the nanotubes formed into a wire; however, in order to transfer between nanotubes, electrons use a method called *quantum tunneling*. This jump in space is something like riding along the highway in a car and suddenly finding yourself in a car in the next lane. This type of wire is being called *armchair quantum wire* because of the use of armchair nanotubes and the quantum nature of electron transport within the wire. If armchair quantum wire becomes available, its first application would be in situations where reducing weight is critical, such as in satellites and airplanes.

There's also the possibility of storing hydrogen in nanotubes. Imagine a material that can absorb hydrogen like a sponge absorbs water. This material could be used as a fuel tank for hydrogen fuel cell-powered cars. (We discuss various research efforts using nanotechnology to improve hydrogen production and storage in Chapter 9.)

Researchers are looking at using nanotubes, as well as other nano-size materials, to make transistors, memory cells, and wires. As the complexity of computer chips increases, it's very important to reduce the size of the devices and wires. Currently the narrowest device or wire used in computer chips is 90 nanometers. With nanotubes and nanowires (see the next section) we could produce transistors and memory devices about a nanometer wide. This effort is called molecular electronics and we talk about it in more detail in Chapter 8.

Wire made from nanotubes is not the same as nanowire. *Nanowire* is a nanoscopic solid wire made of various metals — too small to be seen by the human eye. Wire made *from* nanotubes is like a woven fiber made from many, many nanotubes — and that could be as big as a standard electrical cord.

Getting Wired with Nanowires

Nanowires are simply *very* tiny wires. They are composed of metals such as silver, gold, or iron, or semiconductors such as silicon, zinc oxide, and germanium. Nanoparticles are used to create these little nanowires, which can have a diameter as small as 3 nanometers.

Growing nanowires

The production of nanowires is similar to nanotubes; it requires using a catalyst particle in a heated reaction chamber. To grow nanowires composed of gallium-nitride, researchers at Harvard University flow nitrogen gas and vaporized gallium through the reaction chamber containing an iron target. Iron nanoparticles are vaporized from the target by a laser to act as a catalyst. Both gallium and nitrogen molecules dissolve in the iron nanoparticle. When you get so much gallium and nitrogen in the particle that it starts to sweat off of the surface, molecules precipitate onto the surface of the particle where they combine to grow the nanowire.

When you grow any nanowire, the materials you use must be soluble in the catalyst nanoparticle. For example, to grow silicon nanowires, a gold catalyst nanoparticle is used because silicon dissolves in gold.

To grow arrays of nanowires — great for making electronic devices or sensors — you can use catalyst nanoparticles positioned on a solid substrate, rather than nanoparticles in a vapor. For example, researchers at the National Institute of Standards and Technology have used gold nanoparticles on a sapphire surface as catalysts to grow arrays of nanowires composed of zinc oxide. By changing the size of the gold nanoparticles, they are able to control whether the nanowires grow tilted vertically at a 60-degree angle up from the surface, or horizontally along the surface.

Nanowires at work

Several research groups have demonstrated the use of nanowires to create memory devices and transistors. Researchers at Hewlett-Packard and the University of California at Los Angeles have demonstrated that a memory cell can be formed at the intersection of two nanowires. You can see a crossbar array of these nanowires in Figure 4-7 (courtesy of Hewlett-Packard). Using a somewhat more complicated array of nanowires, they have also come up with a transistor-like device called a crossbar latch.

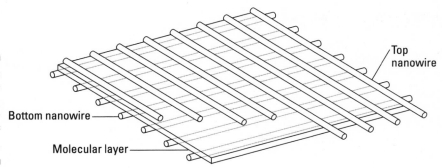

Figure 4-7:
Memory device using a crossbar array of nanowires.

Top nanowire

Bottom nanowire

Molecular layer

Folks at the University of Southern California and the NASA Ames Research Center have demonstrated a memory device that uses indium oxide nanowires. They are predicting that this device will be able to store 40 gigabits per square centimeter, which is a lot of data by anybody's standards.

Building transistors and memory devices used in computer chips from materials about the width of a nanometer, such as nanowires, is called *molecular electronics*. We discuss molecular electronics in more detail in Chapter 8.

Meanwhile, over at Harvard University they've demonstrated a nanowire-based sensor that can detect diseases in blood samples. The working part of the sensor is a nanowire that has been functionalized by attaching certain nucleic acid molecules to it. The nucleic acid molecules bond to a cystic fibrosis gene if it is present in a blood sample. When this happens, the conductance of the nanowires changes. The change in the nanowire conductance causes a current to flow.

This type of sensor has the potential to provide immediate analysis of blood samples for a variety of diseases, possibly right in your doctor's office with just a pinprick in your finger. That's much more convenient than giving vials full of blood and waiting for a test to come back from a lab. Add to that, this sensor is highly sensitive and might detect diseases we've never even been able to detect before, or detect viruses at an earlier stage.

But there's a major challenge for researchers developing this technique, either with nanowires or nanotubes: They have to find a way to make the sensors selective and prevent false signals. In the Harvard demonstration, they did this by using a specific nucleic acid that would only bond to the cystic fibrosis gene. We talk more about nanotechnology in medical diagnosis in Chapter 10.

Finally, researchers at the National Institute of Standards and Technology, as well as the folks at the Max Planck Institute, are investigating the use of nanowires to increase the density of a magnetic recording medium (such as the disk drives used in computers). Both groups have been able to deposit arrays of magnetic nanowires — and their work shows that it's feasible to use this type of structure to store information at a much higher density than current disk drives can. However, other researchers are investigating the idea of using certain arrangements of nano*particles* to do the same thing as nanowires. It's a toss-up as to which idea will win out.

Chapter 5

Adding Strength with Composites

In This Chapter

▶ Utilizing plastics to make lightweight, conducting composites

▶ Creating a stronger fiber to support the materials we use

▶ Coming up with composites that react to their environment

*H*ear the term "composite" and you probably think either "high-priced golf clubs" or — if you're a Michael Crichton fan — airframe structures. While high-priced golf clubs and airframe structures are both intimately tied up with composite technologies, composites have a wide variety of applications that will continue to grow with nanotechnology. It should come as no surprise, then, that we decided to devote an entire chapter to the role composites play in the nanotechnology story.

Compose This!

A composite is an engineered material "composed" of two or more components. There are many types of composites — some are made from flakes pressed and glued together (particle board), others are constructed from fibers (fiberglass). Usually, a fiber acts as a backbone for the material and a matrix binds the fibers together. Fibers are chosen for their strength and the matrix is chosen for its ability to bond to these fibers. The reason we add the fiber is to enhance the strength of the finished material.

Need a handy metaphor? Think of pizza, that manna from heaven for students everywhere, as a loose form of a composite. The bread is the substrate (structural base), the meats are the fiber, and the cheese, when melted, binds the meats together in a matrix. (Okay, maybe it's not the best building material for, say, suspension bridges, but it sure tastes pretty good.) A common composite that you may have heard of — more structurally sound than pizza — is the particle board found in houses and furniture. Particleboard is a plywood substitute that consists of wood flakes and sawdust bonded together with glue and pressed into sheets. Pretty good for a twentieth-century innovation, but nano can do better.

Lighter, stronger, cheaper

The nice thing about composites is not just that they're strong — they are also "moldable." Why use the bulk material (say, aluminum) when a composite composed of glass and epoxy will be just as strong or even stronger? Not to mention the fact that molding bulk aluminum into desired shapes (air frames or boat hulls) may not be as effective as molding a plastic matrix-based composite. This composite may even be lighter and stronger. You can see that composites are our first attempt to make the perfect material — lighter, stronger, and cheaper.

Interfacing the fiber with the matrix

The most important concept to pull away from composites involves the importance of the fiber/matrix *interface* — that place where two unlike materials join together to become (hopefully) indissolubly one. For what good would fiberglass be if the interface could not stand up to the pressures of the job — say, if the fibers easily became disengaged from the matrix? You might be able to market such fiberglass as cheap insulation, but forget about using it in a vaulting pole or as a boat hull.

So, the better your interface is at keeping the bonds between fiber and matrix indissoluble, the greater the structural integrity of the entire composite. We *reaallllyy* want the fibers and the matrix to structurally adhere to each other. (Otherwise, we'd have this sloppy mess that would pull apart like cotton — or, yeah, cheap insulation.) But better interfaces also mean better (more effective) load transfer — transfer of load from the bulk composite to the matrix to the fiber.

And how do we get better interfaces? Funny you should ask. As you might expect, given the title of this book, we're pretty sure that nanotechnology holds the key. We're willing to bet the farm that working at the nano scale will allow us to enhance both the structural integrity of the entire composite as well as the load transfer between both the fiber and matrix, thus adding strength to composites.

One Word: Plastics

When Rich went off to college in 1995, his dad said, "One word: World Wide Web." "But that's three words," Rich said. His dad was, in his own way, trying to express the future importance of the Internet by mirroring the famous line from the 1967 movie *The Graduate,* "Just one word: plastics." At 18, Rich (like most teenagers at that age) thought he knew everything and didn't listen to the advice the way he probably should have. When he graduated from college, his dad approached him again and said, "One word: nanotechnology." This time, he listened.

Plastics (polymers) have gradually infiltrated our lives over the last hundred years, from shiny handles to credit cards to kitchenware to computer cases to cheap furniture. These days they're so commonplace that it's hard to get excited about them — unless you take a nano-perspective. Only then can you see a polymer in all its quirky glory, spilling out as a long, elaborate chain of carbon- or silicon-based molecules (as shown in Figure 5-1) — now not only common but outrageously tweakable at the molecular level.

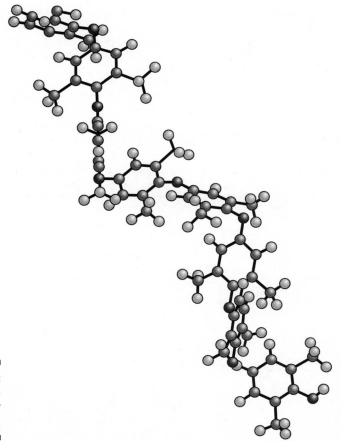

Figure 5-1:
An example
of a polymer
chain.

If you take a look at plastic's family tree, you'll see that plastic's great-great-great-granddad was natural rubber. A momentous year in this ancestor's life was 1839, when Charles Goodyear "vulcanized" rubber by cooking the fellow in sulfur. Up until that time, rubber would become sticky in hot temperatures and brittle in cold temperatures. (Kind of like Rich's own great-grandfather.) Without vulcanization, we wouldn't have adequate tires for our cars, a state of affairs that would have drastically hindered our industrialization.

Another landmark date in the march toward plastic perfection was 1894, when a natural polymer, cellulose, was used to replace expensive silk. (Okay, "rayon" — the cousin of cellulose — didn't have quite the panache of silk, but it was a heck of a lot cheaper and easier to take care of.) "Bakelite," the starting point for umpteen thousands of pieces of costume jewelry, electrical insulators, and radio sets, came about in 1909 and is considered the first purely synthetic plastic and "thermoset" — tough and temperature-resistant. Thermoplastics can be molded, allowed to cool, and then melted again. On the other hand, thermo*sets* are molded once and "cured" — the plastic forms a tight matrix that cannot be undone without destroying the plastic itself. Bakelite — cheap, strong, and durable — shows up in radios (see above), telephones, clocks, and even billiard balls.

Plastics continued to develop through the world wars, throwing off products like Styrofoam, PVC, nylon, acrylic, epoxy, Mylar, Teflon, and Kevlar. Table 5-1 gives a listing of the approximate birth years of some common plastics, as well as their many uses (and a sense of just how fast they've emerged).

Table 5-1	Common Plastics		
Year	*Name*	*Properties*	*Applications*
1834	Vulcanized rubber	Temperature-sensitive	Car tires
Post-WWI	Polystyrene (PS)	Brittle, "foamable"	Eating utensils, Styrofoam
Post-WWI	Polyvinyl chloride (PVC)	Heat/weather-resistant	Plumbing pipes
1927	Nylon	First synthetic fiber	Stockings, parachutes
1933	Polyethylene (PE)	Cheap, durable	Tupperware
1938	Teflon	Corrosion-resistant, low friction	Non-stick frying pans, Gore-Tex jackets
1939	Epoxy	Adhesive	Glue
1941	Polyester	Synthetic fiber	1970s clothing

Year	Name	Properties	Applications
1960s	Kevlar	Strong, lightweight	Bulletproof vests
1970s	Polycarbonate	Strong	CDs, eyeglasses

Dissipating static electricity

The history of plastics is truly a fascinating topic, but we can understand that you may feel that a little bit of plastics history goes a long way. Time to move on to a topic more relevant to the great promise we see on the nanotechnological horizon. That's right, it's time to talk about laundry.

Anyone who regularly pulls their socks out of their dryer has probably had a fateful encounter with static electricity at least once in their lives — zzzzzttttt!! These small-but-startling electric shocks are a handy illustration of electrostatics — the force exerted by an unchanging electric field upon a charged object.

Now, getting shocked by your socks or having your hair stand on end after taking off a sweater may sound pretty harmless, but the truth is, static electricity and electromagnetic interference play a huge role in destroying our electrical circuits and computer equipment. (Forty percent of all failures can be attributed to electrostatic discharge!) Case in point: your computer. If you've ever bought a sound card for your computer, it was more than likely wrapped in pretty silver plastic. These plastics dissipate any static electricity that may build up, protecting the electronic equipment from a sudden surge. And then, when you go to install the sound card, you'll find that Rule Number One in the installation manual accompanying the sound card states that, after you open up the case to access your computer's innards, you have to ground yourself by touching the metal casing. This removes any static electricity that may have accumulated while you were rubbing your socks against the carpet. Ignore Rule Number One and you may end up with a sound card that's been fried by the same little spark of static electricity that shocks you from your socks.

Your home computer isn't the only piece of electronics out there that needs to be shielded. Just ask the U.S. military. They shield their equipment from electromagnetic pulses — for example, those given off during a nuclear explosion (which would severely scramble the brains of any piece of electronic equipment not so shielded). And then there's the Federal Aviation Administration — very much interested in keeping an increasingly automated and electronics-dependent commercial air fleet safe by shielding onboard electronic equipment from interference. Hopefully, the nuclear Armageddon scenario doesn't play out, but you can understand the broad array of uses for protecting electronic equipment — from your home to your transportation to your defense.

It turns out that carbon nanotubes — the same guys who play a starring role in our discussion of molecular architectures in Chapter 4 — also do a fantastic job of dissipating static electricity while dispersed in plastics. How fantastic? Well, the following description of work done by Zyvex Corporation — in conjunction with Michigan Tech — may give you an idea. This work offers a particularly telling look at the difficulties of working at the nano scale — as well as the big payoffs that can be achieved.

The carbon nanotube connection

The folks at Zyvex Corporation saw clearly that one key to an effective composite is the even dispersion of the fiber in the matrix. You don't want the fibers to bunch up just in one area; that would tend to weaken other areas. Unfortunately, carbon nanotubes on their own tend to want to clump together — a direct result of their van der Waals' forces. (No, that's *not* some Dutch prog-rock group from the 1970s.) Van der Waals' forces are a lot like the static electricity binding your socks together when they come out of the dryer — very weak attractive forces. They're different from the chemical bonds that actually attach molecules to each other — those are more like sewing your socks together.

Okay, even dispersion of fiber is important if you want to work with carbon nanotubes — and if you want to keep the individual tubes from bunching up, you have a couple of possible approaches:

- ✔ **Coating the tubes:** The nanotubes are typically *functionalized* (where you covalently bond molecules called functional groups to the side walls, allowing nanotubes to perform a desired function such as to disperse better or attach to other large molecules), but applying this type of coating does pose some risks — the nanotube's unique structural and electrical properties are sometimes negatively affected.

- ✔ **Sonication:** Here you suspend the nanotubes in a polymer solution and bombard them with sound waves. That gives you the needed dispersion, but lengthy sonication can damage the tube walls as well.

Zyvex Corporation found that wrapping the nanotubes with a polymer was a pretty effective way to make sure the tubes dispersed evenly. For example, you can use Poly(aryleneethynylene) molecules (PPEs), which do not wrap completely around the nanotube; instead, they sit on one side, providing what looks like a backbone (as you can see in Figure 5-2). PPE is a rigid polymer that does not bend easily — it sits along the length of the nanotube. Here's where the "static cling" analogy comes into play: A weak, *noncovalent* force (not a chemical bond — no electrons are shared) sticks the polymer to the nanotubes, which keeps the tubes from bunching together. This *noncovalent* functionalization between tube and coating allows the nanotubes to disperse evenly in a polymer without giving up their unique properties, like the previously mentioned *covalent* functionalization sometimes does. Pretty slick (so to speak). As an added bonus, PPE by itself provides unique physical, optical, and electronic properties — stuff that can only help our nanotube.

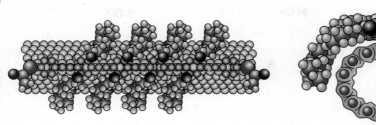

By adding PPEs to the mix, you end up with *single-walled nanotubes* (*SWNTs*), noncovalently functionalized so that each tube acts as a unit that reinforces the fiber. You can then mix this PPE-SWNT *in situ* with a polymer matrix — in this case, polystyrene, the well-known (inexpensive) hard plastic you find molded into plastic drinking glasses and computer cases. What you get is considerably strengthened plastic.

In situ is Latin for "in place," meaning all components are thrown together and mixed as opposed to some composites being added one step at a time or as part of a layering process. Given the nanotube's small size, you can probably guess the difficulty in isolating and combining them one at a time.

Percolate, percolate, percolate

We've got ourselves our own little PPE-SWNT building block (kindly provided to us by Zyvex Corporation) just waiting for us to put it to use. Before we do that, however, we're going to take a coffee break — or at least get a little taste of percolation theory. You may think of percolation when your morning cup of coffee is arguing with the overwhelming need for an afternoon nap — but in mathematics, percolation theory describes a *connected network made from a random arrangement.*

A what and a what? Imagine, if you will, that you have a square piece of plastic and you scoop out holes in the plastic following a fixed pattern. Each hole is (potentially) electrically connected to every other hole. Now, you randomly drop metal into the holes until you get a connecting network — a *spanning cluster* — from one end of the square to the other. This critical point — the point, in this example, where electric current spans from one end of the plastic square to the other — is called the *percolation threshold:* the ratio between how many metal drops are needed to form a connecting network and the total number of holes in the plastic. It's expressed as a percentage; the higher the percentage, the likelier the electrical connection — but the lower the percentage, the less material needed. Figure 5-3 shows the critical route from start (S) to finish (F), indicating that 150 random metal drops, landing in 150 of 253 possible holes (in other words, 59.29 percent of all possible holes) were enough to get a connection from one side to the other. The percolation threshold for this two-dimensional example is therefore 59.29 percent.

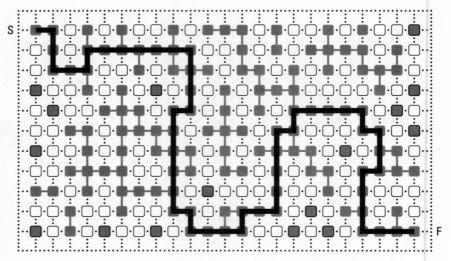

Figure 5-3:
A two-dimensional example of a percolation threshold.

Percolation theory is not limited to two dimensions. Think of a cube of plastic with metal shavings suspended inside. Some of these metal shavings touch; others don't. A high concentration of metal shavings gives you a greater chance of a conductive connection between opposing sides of a cube; lower concentrations of metal shavings reduce that chance. But we're talking efficiency here: What we're looking for is the lowest possible amount of metal shavings that still provides a reliably connected network.

Enter our heroes — the noncovalently functionalized PPE-SWNTs. (Snappy name, isn't it?)

The folks at Zyvex and Michigan Tech found that by mixing these polymer-wrapped nanotubes with polystyrene, they came up with a lightweight, highly conductive plastic. Between 0.05 and 0.1 weight percent of SWNTs (roughly 1 gram of SWNTs for every 1000 grams of composite) produced an incredibly low percolation threshold. Result: a conductive, polymer-based composite utilizing hardly a smidgen of SWNTs.

The use of precious materials (SWNTs) went down; the conductivity went way up. If 7 percent of the material's weight is functionalized (wrapped) SWNTs, you get 14 orders of magnitude (10^{14}) — the conductivity of pure polystyrene — and 5 orders of magnitude (10^{5}) — the conductivity of a SWNT/polystyrene composite that doesn't use PPE to wrap the nanotubes. Clearly, adding PPE to the SWNTs markedly increased their ability to conduct.

Not convinced? Check out Figure 5-4, which shows the rapid rise in conductivity versus the SWNT mass fraction in percentage. This rapid rise in the percolation threshold shows that noncovalently functionalizing the SWNTs with PPE increased the SWNT's ability to disperse — which in turn gives the material two new advantages:

- ✔ Better dissipation of electrostatic charges
- ✔ Electromagnetic shielding capabilities

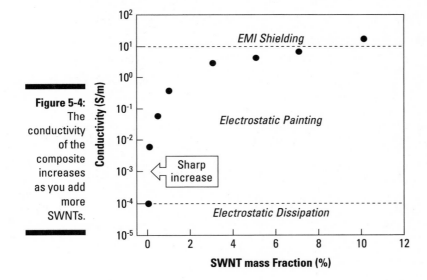

Figure 5-4: The conductivity of the composite increases as you add more SWNTs.

Now for a real-world application: a better paint job. Electrostatic painting makes paint adhere better to metals and plastics by first electrostatically charging the paint particles and then grounding the part to be painted. When applied, the paint particles are drawn in (electrostatically) to the part to be painted — resulting in an even, cost-effective coat (little waste, not much overcoat), which is very important to the automobile industry.

Dispersion is key to achieving good electrical conductivity with a low percolation threshold. Noncovalently functionalizing the SWNTs with PPE aided in this dispersion, and since we didn't have to pile on the SWNTs to achieve the necessary conductivity levels, we don't disturb the host polymer's other preferred physical properties (including how easily it can be processed). All in all, it's a basic improvement to a common (but very useful) process — a very good thing, and a typical nano-solution.

The next example uses the same concept — nanotubes dispersed in a polymer — to emit light and display images.

Displaying images

There will come a time when flat-screen TVs will seem so last-year. (Next year, perhaps?) New advances in display technology — whether for TVs, cell-phone displays, or computer monitors — seem to be arriving every few months. And many of those advances are being driven by two benefits nano-technology promises to provide — lower energy consumption and higher picture resolution.

You may be asking yourself, "Why do I need such a high-resolution picture for watching my TV? I don't want my friends to think I'm a technogeek." Although the entertainment industry can certainly benefit from sharper resolutions — it's what a lot of consumers want — the medical industry will need it most. A doctor's prognosis is only as good as the equipment on which he views your magnetic resonance image (MRI) — and there is certainly a lot of consumer demand for correct medical prognoses.

Questioner that you are, you may be asking yourself, "Why do I need my dis-play to be energy efficient?" You may have noticed that your PDA and cell-phone displays go into a dark, standby mode — this is to save energy. It'd be nice if the display never went off — and you'd still only have to charge your cellphone once a month. You may ask yourself, "Why do I need my display to be lightweight?" Well, okay, maybe you won't be asking yourself that question — but it does bring up a good point that may not be obvious: The U.S. soldier lugs around 80 to 100 pounds of equipment — and they *love* light-weight stuff. (Think about that next time you heft your 5-pound laptop.)

The Korean connection

So who out there in the real world is set to answer consumer demand for high-resolution, energy-efficient, lightweight display equipment? How about Samsung? The Korean electronics giant — prominent as a maker of display screens — has incorporated carbon nanotubes into a working color screen. As in a traditional television (the big, bulky kinds), the carbon nanotubes shoot electrons at phosphors on a glass screen. (A *phosphor* is any material that emits visible light when exposed to radiation — the red, green, and blue colors that you see on your screen.) Unlike traditional TV picture tubes, Samsung's color screens are thin, lightweight, and don't use much power.

In addition to the Korean giant, we want to highlight two smaller companies that are utilizing nanotechnology for displays — in this case, electronic paper: NTERA (which uses electrochromatic technology) and E Ink (which is taking the microcapsules route). The next section spells out their two differ-ent approaches.

The smarter the window . . .

It's pretty easy to come up with a great marketing phrase for describing a potential application of electrochromatics. In fact, here it is on a silver platter: smart windows. Imagine living in Houston, Texas, with bright summer sunlight streaming into your high-rise condo. To "turn down the sun," you flip a switch and your windows go from clear to gray to dark, allowing just enough sunlight in to make your living room a comfortable environment — all the light with little heat.

With that comfy image in mind, here's how it all works: *Electrochromatics* are materials that change color when energized by an electrical current. Electricity starts a chemical reaction which then changes the properties of the material — so it either reflects or absorbs light. The electronics are sandwiched into your glass window, as in Figure 5-5. At left in Figure 5-5, you start off with light shining through the layers of glass. When you turn on the electricity — prompted, perhaps, by the effects of a blazing Texan sun — a reaction occurs and ions move from the ion-storage layer, through the ion-conducting layer, and into the electrochromatic layer — and the glass turns opaque (as shown at right in Figure 5-5). Turning off the electricity has the reverse effect — the window becomes transparent again.

One of the nice things about using electrochromatics is that it's tunable — you're not stuck with just light or dark; you can come up with whatever shade of gray you want, simply by changing the amount of current you use.

Now, the setup you see in Figure 5-5 shows a window kept opaque by constantly running electrical energy through it. Nice, but not the most energy-efficient solution to the problem. This has led to electrochromatic window designs that only require electricity to drive the initial change — changing the window from light to dark or dark to light. Clearly (so to speak), this is the route to go — lower power bills, less hassle, more efficiency. You know — progress.

NTERA, an Irish company spawned from Dublin's University College, figured all this out and came up with a solution they call NanoChromatic Displays (NCDs, for short). The NCD is based on the same concept as the electrochromatics just described — but it goes the extra mile into nanotech territory. (They don't call it NanoChromatic for nothing!) Given their nano size, they have a large number of particles that can quickly switch on or off — important for video and computing — but the display is also *bi-stable:* Each pixel stays colored until actively turned off — no need for backlighting. Fast switching and no backlighting mean low power consumption. The display is easy to see from nearly any angle, and — here's a blast from the past — it uses titanium dioxide (the chemical that makes paper white) to ensure good contrast.

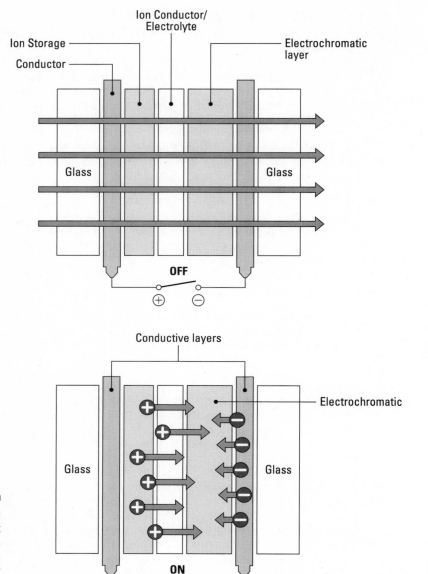

Figure 5-5:
Smart
windows
can change
their
opacity.

E Ink, a company based in Cambridge, Massachusetts (Can you say, "MIT?"),
takes a different approach to electronic ink. Rather than touting nanochro-
matics, they've placed their bet on *microcapsules*. Figure 5-6 gives you
the thumbnail sketch of what E Ink came up with. Each microcapsule has
positively charged white particles and negatively charged black particles

sandwiched between two electrodes. Depending on the charge applied to the bottom layer of the particle, you can get the microcapsule to represent a black (-), gray (-/+), or white dot (+). Zap the layers with the appropriate electric charge and you come up with black text on a white background (or even a grayscale image).

E Ink also touts the key buzzwords — high contrast, low power. However, their microcapsules suffer from slow switching speeds that may limit their application breadth. NTERA may also have an edge on E Ink's production methods — they can utilize existing LCD manufacturing plants and equipment, whereas E Ink may have to start with a custom-designed plant.

Cross Section of Electronic-Ink Microcapsules

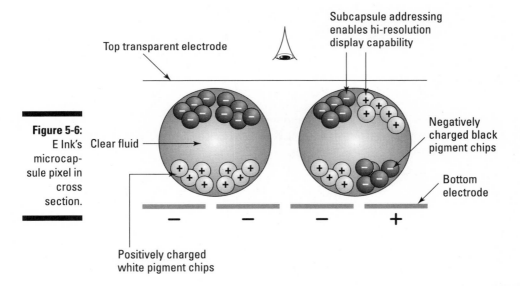

Figure 5-6: E Ink's microcapsule pixel in cross section.

Top transparent electrode

Subcapsule addressing enables hi-resolution display capability

Clear fluid

Negatively charged black pigment chips

Bottom electrode

Positively charged white pigment chips

Lightening the Load with Nanofibers

Fibers can be used as a backbone to many composites. Now, we've been touting the benefits in this chapter of making composites out of nano-scale material, but it's time to let you in on a little secret — when we talk about making nanofibers, we're not talking about some radical new process destined to win the two of us our own personal Nobel Prizes (though we can dream). The processing methods we describe in this section have been used before — they're time-tested. Not that we aren't proud of these processes. They're way cool — carbon nanotube fibers are especially close to Rich's heart (they're the main focus of his research at Rice University). Beyond cool, however, is in how they show reliable, known methods (that we know work well) being put to new ends. Read on.

Nanotubes

No one nowadays can argue with the fact that carbon nanotubes hold great promise in the making of materials with superior mechanical, electrical, and thermal properties. The trick is getting this promise to realize itself as products that better our lives — and provide a return for those who invested in the promise. One method of finessing this trick — namely, of capitalizing on the superior mechanical, electrical, and thermal properties of carbon nanotubes — is to use them to hot-rod some materials. For example, you can form a composite by spinning the carbon nanotubes into a fiber that is then inserted into a matrix. (The "Compose This!" section, earlier in this chapter, has more on the mechanics of creating composites.)

Creating a spun fiber to be used in a composite is not new: body armor used by the military has long been comprised of Kevlar, a polymer-based fiber; airframes and sports equipment utilize carbon fiber (and carbon makes a tough, flexible, useful fiber). The uses of carbon fiber have been increasing whereas the cost is decreasing — which in turn leads to even more uses for carbon fiber. (Figure 5-7 illustrates, in graphic terms, this perversion of the Law of Supply and Demand — here, the more the demand, the lower the price.) Initially, in the 1970s, only the aerospace industry capitalized on carbon fiber, but this trend moved into the sporting-goods industry in the mid-'80s and continues into the industrial and automotive industries. Carbon fibers may also play an important role in energy storage (fuel cells) and the generation of power (wind turbines).

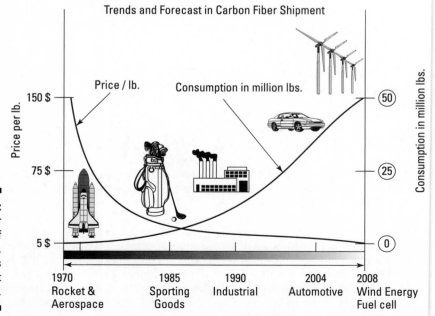

Figure 5-7:
Timeline for the use of carbon fiber. As use goes up, cost goes down.

That's all fine and dandy for traditional carbon fibers, but will a carbon *nano-tube* fiber show similar trends? That will all depend upon whether carbon nanotube fibers are seen as an economically viable alternative to current fibers. If a carbon nanotube fiber has good conductive and thermal properties, has 100 times the mechanical strength of carbon fiber, *and you can produce it at a comparable production cost*, then carbon nanotube fiber is sure to replace carbon fiber for particular applications.

We've been stressing strength as a major benefit of carbon nanotube fibers, but the most exciting applications may not necessarily need raw strength. Coming up with an incredibly strong, lightweight conductive wire would greatly impact energy distribution, for example. (Chapter 9 has more on carbon nanotubes and energy uses.) "Armchair quantum wire" — again, see Chapter 4 — could be used to shuttle electrons around with little resistance, an application which, like carbon fiber, is sure to see its first applications with space vehicles. (After all, lugging all that copper — the traditional material for low-resistance wire — up into space gets pretty expensive.) Clearly, finding a lightweight, durable, energy-efficient alternative to traditional wires not only would help our space endeavors, but also could boost the creation of more fuel-efficient cars and other everyday items.

We think the sky's the limit here, and we want to share our enthusiasm. In the next section, we walk you through one challenge of developing a carbon nanotube fiber.

What a tangled web we weave

If you flip over to Chapter 4, you'll see considerable page real estate devoted to ways people have divined for coming up with carbon nanotubes. Such methods result in a black, fluffy material — the nanotubes have intertwined and knotted together. (Kind of like that ball of twine that always gets unraveled in the kitchen drawer and takes forever to untangle.)

Tangled knots of stuff definitely have their uses — we're partial to fettuccine Alfredo ourselves — but a major goal of nanotechnology is to use existing industrial techniques to manipulate atoms and molecules in bulk to make a large, atomically perfect, usable structure. And tangled knots of stuff do not always produce the most usable structures. This fact has led some to conclude that the first order of business, if you're working with carbon nanotubes, is *alignment* — getting all those nanotubes to line up in the same direction.

Why is alignment, of all things, number one on anyone's list of things to get done? Imagine the following scenario: If you take a plastic drinking straw and hold it at the ends, you can pull and push on the ends of the straw all you want, but it's going to take some real effort to get the straw to deform. However, if you take your finger and poke the straw from the side, the straw easily buckles and bends. Now, the same thing is going to happen if you have a bunch of aligned straws — the force needed to stretch or compress the group of straws is quite considerable compared to the force needed to simply bend the group of straws. Bottom line: Aligned tubes are stronger along their length.

Now imagine carbon nanotubes, our "nanostraws," aligned in a fiber. The fiber will have great strength when you pull — it's like trying to pull those bundle of straws but instead you're pulling carbon nanotubes. This fiber will not be rigid and flexible, so compression and bending at the macro scale won't be a factor causing it to break. Our main structural benefit is gained when pulling the fiber. (If you take a look at Chapter 2, we go through the development of Kevlar, a polymer based fiber. We point out that the polymer molecules are aligned in the same direction — same goes for our carbon nanotubes. This is an example of using a time-tested approach to continue to develop and manipulate at the nano scale.)

Alignment is also important for our "Armchair Quantum Wire." With aligned carbon nanotubes, we can shuttle electrons down the center of them with very little resistance, making our very conductive wire possible.

Getting all our nanofiber ducks in a row means that we end up with a fiber that has a ridiculously high tensile strength. "How high?" you might ask. Check out Table 5-2, which gives the tensile strength and density for common materials in increasing order, from wimpy to muscle-bound. (Note that even though glass fiber is "stronger" than Kevlar, it weighs one gram more per cubic centimeter. Our goal is a fiber that is both strong and lightweight, which is why Kevlar is so popular.) Now compare the carbon-nanotube material — it has a *huge* tensile strength and a density comparable to that of carbon fiber. (Okay, these values for the carbon nanotube are theoretical but there's experimental data to support this.)

Tensile strength in Table 5-2 is measured in Pascals (Pa), a unit of pressure. "GPa" is GigaPascal — one billion Pascals. For good measure, we threw in a column comparing the elasticity of the different materials, since elasticity is a property we consider desirable in a fiber. Elasticity here is denoted by Young's modulus — a value that describes how well a fiber can withstand great strain (pulling) and great stress (forces applied at a particular area). Note that carbon nanotubes have a Young's modulus comparable to diamonds (also made of carbon). See Chapter 4 for further explanation of Young's modulus and carbon nanotube's structural properties.

Table 5-2	Comparison of Common Materials		
Material	*Tensile Strength (GPa)*	*Young's Modulus (GPa)*	*Density (g/cm³)*
Wood	0.008	16	0.6
Rubber	0.025	0.05	0.9

Material	Tensile Strength (GPa)	Young's Modulus (GPa)	Density (g/cm³)
Steel	0.4	208	7.8
Diamond	1.2	1140	3.52
Spider Silk	1.34	281	1.3
Kevlar	2.27	124	1.44
Carbon Fiber	2.48	230	2
Glass Fiber	2.53	87	2.5
Carbon Nanotube	200	1000	2

Putting nanofibers to use: Clothes make the man

Body armor — the stuff humans have wrapped around themselves in hopes that it would protect them from the slings and arrows of outrageous fortune — has always taken advantage of new technology. When man first started hunting, they would cover their bodies with leather. As societies formed and fought one another, metal production advanced from bronze to iron and body armor progressed accordingly, so that chain mail and metal plates came into fashion. As weapons got better (in particular, guns), the metal plates got thicker and heavier. Finally, in 1881, Dr. George Emery Goodfellow, after examining a gunned-down man, noted that a bullet had pierced the man's flesh and broken his ribs but had not harmed the silk hand-kerchief the man carried. This led to Goodfellow's study of the case, intriguingly entitled "Notes on the Impenetrability of Silk to Bullets."

Not that we're suggesting silk as the body armor of the future, but you can see the beginning of the story here, one that reached a high point in the late 1960s when DuPont developed Kevlar, a strong polymer fiber currently used in bulletproof vests. Kevlar is five times stronger than steel of the same weight, which doesn't seem to make much sense until one realizes that the key to bulletproof vests is *dispersing the energy* of the bullet over a wide area. This is similar to what happens when you hit a soccer ball into a goal — the net backing absorbs the energy of the ball, dispersing it over a wide area. The Kevlar fibers are also interlaced — when the bullet hits a horizontal fiber, that fiber pulls on every interlaced vertical fiber. Additionally, each vertical fiber continues to push on each subsequent horizontal fiber. This allows the entire sheet of Kevlar to absorb the energy of the bullet, regardless of where the bullet actually hits. (Figure 5-8 shows a cross section of a bulletproof vest. Note the interlaced Kevlar fibers sandwiched together.)

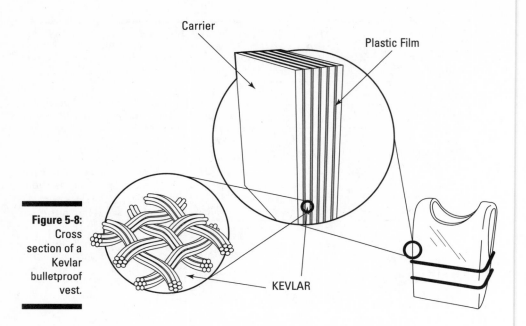

Figure 5-8:
Cross section of a Kevlar bulletproof vest.

Now, as technology progresses, you can expect stronger fibers to emerge replacing Kevlar. Spider silk — yes, spider silk — is considered by many to be an up-and-comer in this field. Although spider silk involves biotechnology more than nanotechnology per se, scientists at Nexia Biotechnologies are breeding genetically engineered goats to produce milk high in spider silk proteins. This is then spun into a fiber — baptized with the name Biosteel — which can be 20 times stronger than an equivalent strand of steel.

Nanotechnology is set to one-up biotechnology with carbon nanotube fibers. Carbon nanotubes promise to be even stronger than Biosteel — 100 times stronger than steel! With a perfect carbon nanotube fiber, chained together with no loss of strength, you could lift an automobile with a piece as thin as sewing thread. That's some strong stuff. And bulletproof vests are not the only applications for carbon nanotubes. Weaving nanotube fibers in with those in existing clothes can become a way to store energy (like a battery), which could then power various electrical devices and sensors to monitor human vital signs. The possibilities are endless.

Speaking of clothes and nanofibers (and, very loosely, a composite), Eddie Bauer and Dockers have embraced Nano-Tex's spill-resistant fabric. You may have seen the commercials with the clumsy guy spilling wine all over his pants, but instead of the wine leaving a nasty stain, it beads and rolls right off, "like water off a duck's back" as Nano-Tex's Web site states. Nano-Tex gives very little insight into *how* they incorporate their spill-resistant technology (something about intellectual property), but we suspect they coat the fibers with a specific polymer. Since the fibers are coated and integrated into the fabric's weave, the cloth maintains its resistance even after many washes — not to mention allowing the fabric to maintain its soft texture and feel. Other technologies that Nano-Tex lays claim to involve a quick-drying cloth (whisks away moisture so you feel cooler) and a cloth that feels just like cotton (but is really a synthetic that maintains color, durability, and wrinkle-resistance but still feels soft). A word to the wise: Be sure that you're actually wearing your "Stain Defender" Dockers before you intentionally spill red wine on yourself while chatting around the buffet table at the office Christmas party in hopes of impressing your co-workers. Rich learned that the hard way.

Clemson University researchers have also integrated a polymer film mixed with silver nanoparticles to produce a "self-cleaning" coating for clothes. This coating is infused into the fabric, creating a series of microscopic bumps that cause dirt to bounce off when water is applied. This is different than other water-repellent materials like Teflon because this film is integrated into the fabric, not applied as an additional layer to the fabric — allowing it to last longer. Also, the particles are too small to be seen since their size is smaller than the wavelength of light. Outdoor items, such as convertible car tops and lawn furniture, are sure to take advantage of this nanomaterial — just imagine being able to simply hose something down and wipe it off. Additionally, the need for less detergent could help our environment.

Put it all together and what do we have? Self-cleaning, wrinkle-resistant, bulletproof clothes that keep us cool in even the most tropical of climes — something Q would be proud to offer James Bond.

Putting nanofibers to use: Into the wild blue yonder

With increasing oil prices, airplane makers are looking to lightweight composites — lighter planes use less fuel getting off the ground, cutting down on operation costs. Airframes currently use carbon fiber for some component parts, greatly reducing the weight. The fiber is woven in much the same way as Kevlar vests (described in the previous section) — with the fiber aligned with the direction of high stress. More or less fiber gets used depending on the amount of stress in a particular direction. The leftmost drawing in Figure 5-9 shows an even distribution of stress.

The central drawing in Figure 5-9 shows more stress along the vertical access, resulting in more fibers in that general direction. The rightmost drawing in Figure 5-9 shows the situation where there is high stress in the vertical and diagonal direction but less in the horizontal direction. All in all, we are optimizing the composite to give the most amount of strength with the least amount of fiber, thus lowering the weight of the entire system. These fibers may one day also sense and respond to stresses — imagine an airplane's wing sensing the stresses upon it and *responding* — by either deforming to reduce stress or becoming more rigid to handle the stress. (The "Sensing strain" section, later in this chapter, addresses these possibilities in a bit more detail.)

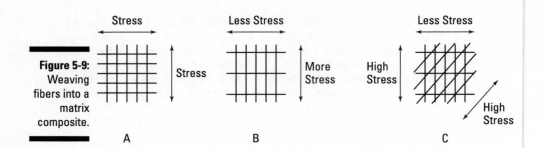

Figure 5-9: Weaving fibers into a matrix composite.

One of Rich's favorite novels, Arthur C. Clarke's *The Fountains of Paradise,* describes a space elevator very much like Jack's beanstalk — an elevator climbing a cable that reaches far into space. In Clarke's novel, the cable supposedly contains a microscopically thin but incredibly strong "hyperfilament" derived from diamond as the main component. Clarke later identified another form of carbon, the buckyball (named after Buckminster Fuller), as the potential hyperfilament.

The space elevator is not a new concept. In 1895, the Russian space visionary Konstantin Tsiolkovsky imagined a 22,000-mile-high tower that would top out at a point in geosynchronous orbit around Earth. A space elevator replaces the tower with a cable; a mass at its end keeps it taut against gravity by revolving around Earth at orbital speed, outside the atmosphere. During the 1960s space race, the Russian Yuri Artsutanov also developed the idea. What *is* new is the proposed main element in the composition of the elevator's track — what material could be strong enough to withstand the winds and debris constantly hitting taut cable that long? This is where the carbon nanotube comes in.

During the summer of 2004, *Discover* magazine and *Scientific American* explored tethers in space. The *Discover* article followed aerospace engineer Brad Edwards's trek towards a feasible space elevator, which had Edwards identifying a carbon nanotube/polymer composite fiber as the backbone of the cable. The cable is set to extend from a refurbished oil-drilling platform to a distance of 62,000 miles out, where it ends in a counterweight. Ascent vehicles would then ride the cable out into space, dropping payload costs to $100 per pound (as opposed to NASA's current $10,000 per pound).

But why confine the benefits to Earth? The *Scientific American* article also illustrates how a tether between two space outposts could give them both artificial gravity. Figure 5-10 shows such a rotating system: Two capsules are attached at opposite ends of a tether and spun; the outward force that pulls the tether taut acts like artificial gravity for the inhabitants. The article also mentions that a tether could be used to slow down decommissioned satellites so they burn more completely as they fall out of orbit — reducing space debris.

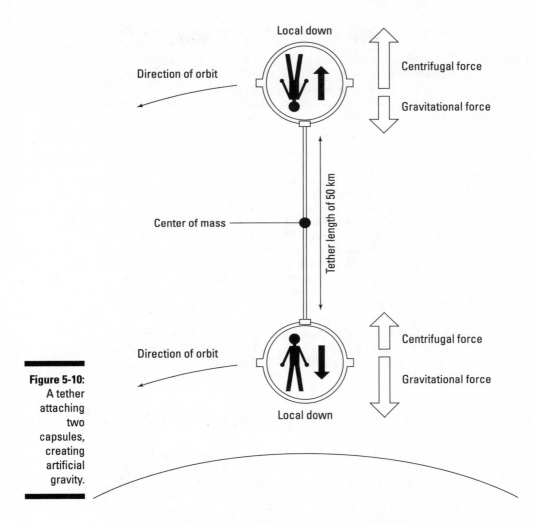

Figure 5-10:
A tether
attaching
two
capsules,
creating
artificial
gravity.

Okay, even though *Discover* and *Scientific American* are reputable journals, we're sure our description here has left you with a few questions. As in:

✔ **Is this feasible?** Maybe. Maybe not. It depends on whether going into space becomes enough of a priority to make developing this space elevator practical. If so, it will be done — and nanotechnology will play a role in its development. Creating a strong enough cable is the crux of a successful space elevator; nanotechnology can create fibers that are stronger and cheaper — probably enough so to withstand the forces that make this concept work. Clarke pinpoints when it can all happen: "About 50 years after everyone stops laughing."

✔ **Is this practical?** Economics will be the driving force. The International Space Station will need 77 metric tons of booster propellant over its ten-year lifespan, at a cost of roughly $1.2 billion, just to keep it in orbit. Decreasing the cost per pound to $100 from $10,000 is a great motivator. Then again, what's so great about space? It's just that, space — empty space (if you ignore all those planets, stars, comets, asteroids, and such that zoom around in it). Now to the next question . . .

✔ **Why do we constantly want to go into space?** The original motivation behind the space race was national security — whoever could deliver a nuclear weapon to an opponent first would win, and the quickest route to your opponent would be through space. Now, corporations have begun to occupy space . . . and we're not talking about Planet Starbucks or the Microsoft Galaxy. Imagery satellites are used for surveying — and some can definitely determine what's *beneath* the earth. (Can you say diamonds, gold, and oil?) The global positioning system (GPS) not only helps land planes but keeps marriages together when the guy refuses to ask for directions. Even telecommunications will continue to rely on space as our global economy continues to grow. Space tourism may be just over the horizon; after SpaceShip One claimed the $10 million Ansari X-Prize in 2004, Sir Richard Branson was impressed enough to "launch" Virgin Galactic, a spinoff from his giant Virgin group of companies. Perhaps solar satellites will beam energy back to Earth. There are plenty of motivations, some identified and others yet to be dreamed. And the dreamers are already busy.

Raising the Bar with Smart Materials

Smart materials are materials that can sense or react to their environments. In this section, we highlight metals and plastics that respond to change by moving, some new sensors that are smaller than ever before, and composites that actually fix themselves — and all this is just the first step. The second step is more along the lines of the liquid-metal robot from *Terminator 2*. Okay, maybe not this year, but it certainly would be cool . . . as long we can program out that inconvenient homicidal-mayhem business.

Coming back to normal

Shape memory composites are materials that retain their original shape no matter how bent out of shape they get. An example of shape memory metal (but not necessarily a type of composite) would be the Flexon eyeglasses utilizing Nitinol, a Nickel (Ni) and Titanium (Ti) alloy developed by the Naval Ordnance Laboratory (NOL) in the early 1960s. (Get it? Ni-Ti-NOL = Nitinol.) This composite has been incorporated into eyeglass frames — even after a user twists them, bends them, or even sits on them, the frames (almost always) come back to their original shape. It's one small example of a commercially viable product that comes from military and NASA research (and there are hundreds). Shape-memory composites come in two flavors: metallic (alloy) and polymer.

Better living through SMAs

Nitinol is an example of what we call a *shape-memory alloy,* or SMA. SMAs have two properties that make them what they are:

- **Pseudo-elasticity:** Is an SMA elastic? Not quite, but sort of. This quality is the rubberlike "give" that every SMA has.

- **Shape memory:** This is an effect of the mechanical energy stored in the SMA when it's reconfigured and cooled. The stored energy is what returns the SMA to its original shape when heated, even if it's been severely deformed.

Now, these SMAs go through what is called a *solid-state* phase change. You may remember phase changes from high-school experiments with water — water changes its phases at different temperatures like so many changes of clothing, going from a solid (ice) to a liquid (water) to a gas (water vapor). SMAs are similar because the molecules do in fact rearrange, but they differ in that their molecules remain closely packed, causing the substance to remain as a solid (thus the qualifying adjective *solid state*).

Figure 5-11 illustrates the difference in molecular structure at each phase as an SMA deforms and reforms. Starting at the lower left, the SMA is in its original form and room temperature state. Applying a load makes the molecules deform. We then remove the load — after adding a bit of heat, the molecules become rigid and form a tight structure. After this tight structure cools naturally and no additional load is applied, the molecules relax, forming a different lattice structure but still maintaining their general molecular orientation and macro-scale physical structure.

SMAs, like many things in life, have their advantages and disadvantages:

- The good news:

    SMAs are biocompatible — human bodies won't reject implants or other materials made of SMAs.

    SMAs have admirable mechanical properties. (In other words, they're strong and corrosion-resistant.)

- The bad news:

    SMAs are expensive to manufacture.

    SMAs have poor fatigue properties — a steel component lasts a hundred load cycles longer under the same loading conditions.

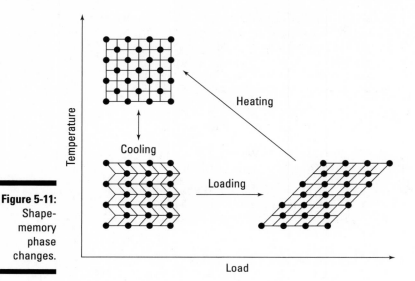

Figure 5-11:
Shape-
memory
phase
changes.

You're probably thinking that SMAs may be great for the eyeglass-frames industry, but what's in it for those of us with 20/20 vision? We can only say that SMAs have already been incorporated into many applications currently in use. If your ears pricked up at the mention of SMA's biocompatibility, you won't be surprised to hear that SMAs are used in many medical applications, including many associated with reconstructive surgery.

As part of our continuing effort to provide real-world examples of why nano-tech matters, imagine if you will the following scenario: You've had a slight altercation with the basement door, one dark and stormy night, and walk away with a busted jaw. You are fully aware that bone plates are used to hold facial bones together during reconstruction, but you are also aware that when a steel plate is used and attached, the bones may initially be correctly compressed to speed up the healing process — but the necessary tension is quickly lost. That's a problem; it means that the fracture is no longer under compression — and that slows down the healing process considerably. A technologically proficient doctor explains to you that when an SMA plate is used instead of steel (as in Figure 5-12), the plate is initially cooled (it's deformable to a certain extent), is attached to the break, and — as body heat raises the temperature of the SMA — it contracts, applying sustained pres-sure to help healing. Now which plate would you opt for: steel or SMA?

We wish this was a done deal, but problems do arise in designing SMA plates to apply the appropriate pressure — for example, how to manipulate the SMA composition so it gives us the "pull" strength required? Well, that's where nanotechnology comes in (manipulating individual atoms in compositions), but more research needs to be done to provide a real and practical solution.

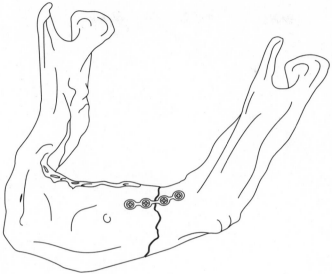

Figure 5-12:
An SMA plate attached to a jawbone.

As for other real-world applications, think aerospace and robotics. Airplane maneuverability depends on moving the flaps at the trailing edge of the wing, as shown on the left side of Figure 5-13. The current way to do this is to use pressure and hydraulics, a process requiring constant maintenance. Using SMAs, we can "warp" the wing instead, as shown on the right side of Figure 5-13, removing the need for hydraulics — which also removes their additional weight and maintenance. SMAs can also be used to mimic human muscles. Instead of waddling around like C3-PO (did you ever see him move his fingers?), robots will utilize SMAs to move their limbs because SMAs are strong, compact, and provide a fluid, lifelike movement. Additional applications include vascular stents (tubing to hold open blood vessels) and even cellphone antennas (bendable yet durable and recoverable).

The SMP connection

We couldn't close this section without at least mentioning SMPs (that's *shape-memory polymers* to you newbies out there). SMPs have their own attractive qualities — they have the capacity to recover from large strains, for example. SMPs operate slightly differently than their metallic cousins. The polymer is raised past a certain temperature, deformed into the desired position, and then allowed to cool. It retains this deformed shape until heat is again reapplied past a specific temperature.

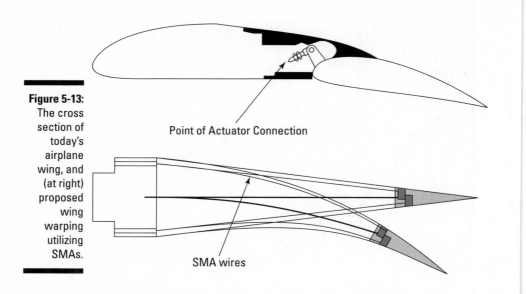

Point of Actuator Connection

SMA wires

Figure 5-13:
The cross
section of
today's
airplane
wing, and
(at right)
proposed
wing
warping
utilizing
SMAs.

There are currently a few downsides to using the polymers over the metals: They are not as strong during constriction, and not as stiff when in place. Figure 5-14 shows an SEM image of two nano-composite shape-memory polymer beams. Researchers at the University of Colorado, in conjunction with Composite Technology Development, Inc., have introduced SiC (silicon and carbon) molecules into the polymer that correct some of these problems. They note that increasing the SiC content till it's 40 percent of the polymer's weight increases the material's recovery strength and recovery time. This is just another example of nano-scale manipulation — controlling molecules and guaranteeing dispersion — strengthening composites at the molecular level. Applications include Biological MicroElectroMechanical Systems (BioMEMS) — therapeutic medical devices for minimally invasive surgery (imagine tiny gadgets gripping and releasing things *inside* blood vessels).

Sensing strain

Sensing strain in structures is important for many reasons. Identifying a problem early in the structure's lifespan could greatly reduce not only the cost and time associated with the repair but could also reduce any potential loss of life from a possible accident. The trick has been figuring out the best way to monitor strain.

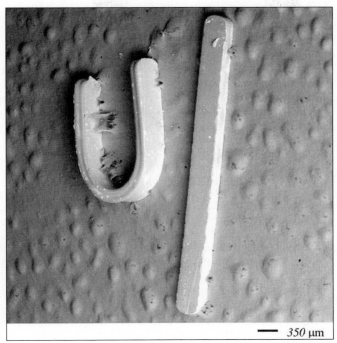

Figure 5-14:
SEM image
of two nano-
composite
SMPs.

$\qquad$ *350 μm*

(Reprinted from Acta Materialia, Vol. 50, Gall et al., "Shape memory polymer nanocomposites," pages 5115-5126, Copyright (2002), with permission from Elsevier.)

There's a multitude of ways to use nanotech to sense strain in a structure — but presently they're mostly restricted to the laboratory. It's easy to measure the load on an airplane wing in a controlled environment, but how do you measure those same stresses when you're out flying in a blizzard? Bridges and airplanes undergo a variety of stresses while doing their jobs, whether from carrying a great load or withstanding various weather conditions. Oil-drilling equipment (yes, back to energy) could use nanosensors to monitor the drill bit in real time as it digs — instantly detecting and reporting the stresses resulting from high temperatures and pressures. In the oil world, the cost of having to stop and repair drilling machinery can be immense — especially for offshore rigs — if a fracture in a pipe is not detected early.

The key is to incorporate the sensor into the actual structure — that way measuring the environment's impact in real time is just another thing the material does (so to speak) automatically. Incorporating nanosensors into a composite is certainly one way to solve this problem. That's great stuff, but think how much greater it would be if the sensor also invoked an action — imagine a bridge that reacted to a greater load by becoming "stronger" where it needed to be, or an airplane that optimized its own shape to reduce drag. Already composites are used in bulletproof vests that harden when they're struck with great pressure.

Ideally, we want our sensors embedded in composites that would not only "sense" but also reinforce our composite. The result would be a material that's not only "smart" but also stronger and lighter. Now, the real trick is not only to get a nano-size sensor incorporated into a composite but also to avoid damaging the integrity of the entire composite itself. We describe (earlier in the chapter) how important the matrix/fiber interface is — and any defect (or foreign substance) within a composite can propagate weaknesses and destroy the system. Some nano-size sensors and concepts are presently possible and worth noting: carbon nanotubes, fiber optics, and superfine magnetic wires.

Carbon copies

The first strain sensor in our potential bag of tricks consists of our old standby, carbon nanotubes. Carbon nanotubes change their electronic properties and resonant frequency when subjected to strains. The easiest way to understand this is to imagine a Chinese finger cuff. As you put in your fingers and pull in opposing directions, the finger cuff gets tighter. The same is true for carbon nanotubes — bending, twisting, or flattening changes their electrical properties — which in turn can be measured. Work done at Rice University affixed "buckypaper" (a disoriented network of carbon nanotubes formed into a flat sheet) to a brass plate. When they stretched the brass plate, they measured the change in voltage across the buckypaper. This voltage change correlated to measurable stresses induced on the brass plate.

Another experiment at the University of Delaware measured the change in resonant frequency of a carbon nanotube. Think of plucking a guitar string and hearing the pitch rise as you stretch the string. Now, if we pull on a carbon nanotube, the frequency at which the nanotube naturally vibrates will change in much the same way. The change tells you something about what's happening to the tube. Imagine, for instance, a bridge that (in effect) says "ouch" when it's under unusual stress.

Both of these approaches show the importance of carbon nanotubes in composites, sensors, and materials that are both composites *and* sensors. (See Chapter 8 on the role nanotubes can play in chemical sensors and Chapter 10 for biological sensors.)

Trip the light fantastic

Another strain sensor we should mention incorporates fiber optics — glass fibers that can send multiple wavelengths of light along them. (Fiber optics have long been used in the telecommunications industry, where they're mass-produced at low cost.) The sensor version of fiber optics — known as a Fiber Bragg Grating (FBG) sensor — enables a material to register stress in real time. Such a material incorporates different grating sizes throughout each fiber. Each grating reflects a specific wavelength of light back to a detector in the light emitter.

This particular grating is like the ones you see over sewer pipes — it's made of slits. Any physical change to these slits (say, by stretching the grating or making it expand or contract with a temperature change) can cause a noticeable change in wavelength. When these wavelengths change, a stress is recorded — which makes this a pretty good system for telling you when a bridge is about to buckle, or that the gale-force winds outside are about to tear the roof off your five-star hotel.

There are three main advantages of the FBG: It requires no electricity, just light along the fiber — no electricity means no possibility of electromagnetic interference by other sensors; the fiber can be incorporated into a composite; and the fiber can be used to measure temperature, pressure, and strain.

Copper: Not just for pennies anymore

We're going to close this section with an old friend — copper wire. It turns out that coating a copper wire with interspersed magnetic coatings creates a wire that can sense strain much the same way the Fiber Bragg Grating does. Figure 5-15 shows a magnetic coating surrounding the copper wire at different increments. The magnetic coating controls the wire's *impedance* — the degree to which the wire resists the flow of electricity — much the same way as a resistor does. As the magnetic coating is stretched, the impedance changes — and the current is measured. This sensor has two advantages over the FBG — small size and low cost — but it does require electricity, and that can mean electromagnetic interference that may impede other systems.

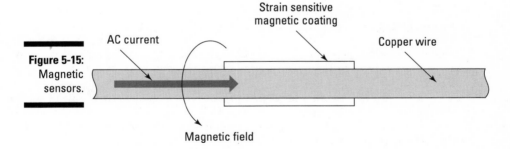

Figure 5-15: Magnetic sensors.

Strain sensitive magnetic coating

AC current

Copper wire

Magnetic field

Heal thyself

If you're familiar with Greek mythology, you already know that even a hero can have a weakness. A composite's Achilles' heel is the interface where the fiber meets the matrix that holds the fiber. If the fiber ever separates from the matrix, microcracks occur inside the composite and begin to spread, weakening the entire structure. (Such microcracks can result from temperature changes or the pressures of mechanical loading.) As you can imagine, it's very difficult to detect and repair cracks within the composite. (You have to call in the big guns — ultrasonics, infrared thermography, and X-ray tomography.) Usually, when a crack is detected, a resin is injected through an access hole to fill the cracks.

Scientists at the University of Illinois at Urbana-Champaign have demonstrated a unique self-healing composite. It involves dispersing microcapsules and catalysts within the composite. When the catalyst comes in contact with the healing agent inside the microcapsule, the healing agent polymerizes and hardens. Figure 5-16 illustrates this process.

Here a crack forms on the left (i), penetrating the microcapsules, which break and fill the crack with the healing agent (ii). As the crack gets bigger, it eventually comes in contact with — and consumes — the catalyst, which then polymerizes the healing agent in the crack (iii). The healing efficiency for the self-healing composite was 50 percent at room temperature and 80 percent at 176°F. This may not be perfect, but the concept shows promise and may become more effective in various ways:

✔ Finding the best percentage of catalyst and microcapsules for the concentration

✔ Increasing the dispersion of healing agent

✔ Attaching the catalyst to the surface of the fiber or the microcapsule itself

At this point, a few sticky questions come to mind:

✔ **Will the microcapsules and catalysts introduce their own flaws, similar to those found in our original fiber/matrix interface?** It's possible — and the interfaces where the matrix meets the microcapsule and the catalyst can also be weak points in the material's structural integrity. Of the two, however, the fiber-meets-matrix interface is the more important one in the composite. It's where the energy of stress is transferred from the composite through the matrix and onto the fiber backbone — a vital point in the structure. The microcapsules and catalysts may not experience the same stress-transfer problem that the fiber/matrix does.

✔ **Will the "polymerized healing agent" also create a flaw?** It could. However, the spherical shape of the microcapsules will disperse the energy. During World War II, when a crack developed on an airplane wing, they'd drill holes at both ends of the crack to prevent the crack from propagating. This disperses the energy so the crack won't get any bigger. The microcapsules' spherical shape acts in much the same way — dispersing the energy over a larger area so it isn't concentrated at a single point. Figure 5-17 shows a cartoonish idea of this concept.

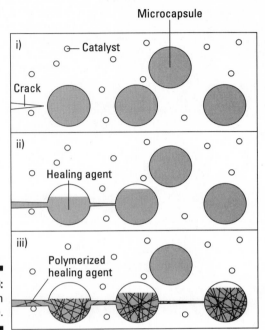

Microcapsule

i) Catalyst

Crack

ii) Healing agent

iii) Polymerized healing agent

Figure 5-16:
Better than
a bandage.

Figure 5-17:
(A) The
crack tends
to go along
the same
direction
if not
stopped. (B)
Introducing
a spherical
hole
disperses
the crack's
forces.

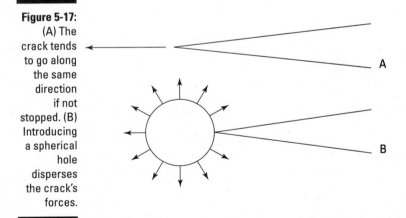

A

B

✔ **If dispersing the energy is only effective half the time, what happens
if the crack grows again?** This is definitely a problem. Composites get
used in high-stress environments where they undergo constant and
repetitive stresses. Therefore, the probability is high that cracks will
occur in the same places more than once. One answer: More — but
smaller — microcapsules. More study is needed to see if this option is
viable for composites in the future.

Part III
"Smarter" Computers! Faster Internet! Cheaper Energy!

The 5th Wave · By Rich Tennant

"We're researching molecular/digital technology that moves massive amounts of information across binary pathways that interact with free-agent programs capable of making decisions and performing logical tasks. We see applications in really high-end doorbells."

In this part...

So exactly what is being done with nanotechnology out in the real world, even as we speak (okay, write)? Lots. Nanotechnology is (or soon will be) making computers more powerful, speeding up the delivery of data, lighting our way, and producing energy.

In this part, we explore uses of nanotechnology in computer memory and semiconductor applications; telecommunications and sharper TV screens; for lighting, and as transistors and sensors. We note some almost-ready advances in producing cleaner energy through nano-tweaked solar cells and fuel cells. And if you've got a pot of cash and are thinking of investing in nano in its next phases, this part will definitely be of interest!

Chapter 6

Building a Better Digital Brain

"I am putting myself to the fullest possible use, which is all I think that any conscious entity can ever hope to do."

HAL 9000, computer from 2001: A Space Odyssey *(1968)*

AL 9000 was the central computer for the spaceship in *2001: A Space Odyssey,* the Stanley Kubrick film based on a story by Arthur C. Clarke. In the language of the movie, HAL supposedly stood for "Heuristically programmed ALgorithmic computer." Heuristics and algorithms are central to defining intelligence: learning by self-discovery (heuristic) and methodical problem-solving (algorithmic). HAL, besides running amok and seizing control of the spacecraft, is a central character in the film, a "conscious entity." Which leads to an intriguing question: How can a computer be conscious — self-aware — hip to its own existence?

Human technology is traditionally a quest to make life easier — from inventing the wheel (to aid mobility) to creating the computer (to solve complicated calculations). Take a look around you and you'll notice that humankind's innovative bent has accelerated of late — particularly in the computer industry. It seems a new computer processor crops up every year that outdates the last one by leaps and bounds. These faster processors may one day create an advanced computer, one capable of working alongside humans, helping us make important decisions — similar to HAL without all the running amok stuff.

To a large extent, the computer's processing ability relies upon the size of the device — the smaller the device, the better. Not only does the smaller size speed up computer processing, it also allows greater portability — after all, where would laptops, PDAs, and cellphones be without fast computing power and small size? As our demands increase for faster and more easily portable computing power, so does our demand for smaller and smaller computer chips. At the same time that we want our processors smaller, we want them to perform more functions and consume less power. But there's a physical limit on how small a chip can be when we use our current top-down manufacturing techniques. Heat dissipation and electronic interference are also big hurdles that prevent us from going smaller.

There's also a practical limit on how much money goes into creating a specialized facility to make those tiny chips. These days, the sticker shock is getting absurdly large, compared to the cost benefit — but traditional methods of mass production simply won't pass muster. New methods of designing computer chips are needed — and are being explored. Of course, if you assume that smaller is definitely better, the next step's almost a no-brainer: Nanotechnology will lead the way to "smarter" computers that will one day have enough tiny, super-capable circuits to "think." Hey, it could make our lives easier — we think.

Linking the Brain with the Computer

It's not as if computers came from some obscure corner of the ninth dimension — their models are actually pretty close to home. If you look at computers just a tad sideways, you'd see that there's a pretty explicit link between the way computers work and the way our brains work. A computer relies on binary symbols — 1s and 0s — because they correspond to the two states of a switch: either on or off. These 1s and 0s can be combined to solve problems — whether they're as complex as the simulation of a nuclear explosion or as mundane as determining the color that a pixel represents on your screen. Our brain contains cells called neurons that store information in the form of electronic pulses and communicate with other neurons in the brain and throughout the body by sending various chemicals. To a certain degree, these electrical pulses (known as *action potentials*) can be thought of as an "on" switch — the electrical equivalent of a 1 in binary. Before we get too excited about such parallels, however, it might be useful to pause for a moment and break down some size comparisons.

This brain/computer comparison is meant to be taken lightly. The brain is certainly more complex, relying as it does on "weighted" signals. In other words, the stronger the connection between two neurons, the more influence (more weight) it has on the resulting calculation. Also, the brain is a bit fuzzy and signals are not necessarily "on" or "off" — there's a lot of gray area in that gray matter. Customarily, computers do their work in precise units of measurement:

✔ **bit (b):** A single *bi*nary dig*it* representing either a 1 or a 0. Transfer rates in computer networks are described in terms of bits per second.

• **megabit (Mbit):** One million bits (1,000,000 bits)

• **gigabit (Gbit):** One billion bits (1,000,000,000 bits)

✔ **byte (B):** Eight bits — referred to when describing units of storage in computers.

   • **megabyte (MB):** One million bytes or eight million bits.

   • **gigabyte (GB):** One billion bytes (1,000,000,000 bytes)

   • **terabyte (TB):** One trillion bytes (1,000,000,000,000 bytes)

We rounded the bits and bytes here for simplicity. Since we're dealing with binary, we use bit-and-byte sizes rounded to the closest power of two. For example, a megabit actually has 2^{20} (or 1,048,576) bits instead of the rounded 1,000,000 bits we just described.

If you've been around any computer at all during the past few years or so, you've heard this bit-and-byte-and-megabyte stuff countless times. But have you recently thought at all about what you've been carrying around inside your brain in terms of "storage units"? Here are the (amazing) facts. There are about 100 billion neurons in a human brain, each with a thousand connections — resulting in 100 trillion connections, all capable of simultaneous calculation — extreme parallel processing, in other words. (*Parallel processing* is where you simultaneously execute the same task on multiple processors.)

To put this in perspective, imagine that one of your friendly *Nanotechnology For Dummies* authors has some laundry to do. (Not too hard to imagine . . . Rich has been a little lazy . . . had a bit of work to do — say, writing a book in his spare time — but now his clothes are beginning to reek a bit . . . bad for his mojo.) Late for his hot date, he takes his laundry and a handful of quarters to the laundromat next door. Rich sorts (yes, he sorts) and distributes his laundry into three machines — one for lights, one for darks, and one for dress clothes. Instead of working at home with one machine and doing each load one after the other, Rich does all three simultaneously — in parallel, if you will. Simultaneously washing (and, later, drying) three loads of clothes at once greatly reduces the time it takes to get the job done.

That good ol' human brain works much like this parallel-processing model, despite the disadvantage of slow-speed neural circuitry. For pattern recognition — a parallel-processing specialty — the human mind does a great job. For extensive sequential thinking, however — having that billionth thought in a row — the poor old gray matter falls short. Easy to see why: Life is short.

And Fast Is Good Because . . . ?

We ultimately want a computer with fantastically fast circuitry (the opposite of our slow neurons) that can work in parallel (like our brain architecture). Fast computers can be used to simulate our world — at both the molecular level and the atmospheric level. In such a world, drugs could be developed and tested on simulated bodies, allowing for catered medicine. Imagine a situation where we program a simulated *you* and a variety of drugs are tested on said simulated *you* until the best drug is found. This drug is then produced and given to . . . you, the real you. Optimata, an Israeli company, already does this by creating a software "clone" of the patient, which helps doctors customize cancer treatment. NEC built the Earth Simulator to simulate the interaction of atmosphere, oceans, and land to predict natural disasters and the effects of environmental destruction. This model aids in understanding how industrialization will affect local rivers and air quality.

On another tack, secure communications and encryption are incredibly important, not only for governments but for large businesses as well — and parallel processors and faster computers developed by other countries/businesses are going to make today's encryption easy to crack. Nanotechnology is set to create better software "locks" to meet this challenge, ensuring security. The emergence of identity theft and the threat of terrorism has contributed to the rise of *biometrics* — identification based on a person's physical characteristics. Of course, for biometrics, you need blazing computing speed to handle the mass of details that describe a unique face or provide immediate voice recognition — and new processors built with nanotech can deliver that power. To make it real (not just something seen in a spy movie), you have to think small — very small.

In the larger world, economic forces are sure to dictate whether dreams of faster processing power become reality. For example, is it cost-effective to build processors with such speed? And how much value does society place on this increased processing capacity? And if we do create artificial intelligence (AI), will we grow bored because we've left it to machines to solve all our problems? Will we be able to control the machines or will they control us? (*The Matrix*, anyone?) Hey, if there's artificial intelligence, will we also have to deal with artificial stupidity (AS), or can we make do with the real enchilada? These are deep questions to ask and something to think about as we build that better mousetrap.

End of the Transistor Road

Computers do their computing in binary, using 1s and 0s to describe . . . anything you can imagine. Now, a *transistor* is the switch that says whether a bit is either on or off — that is, a 1 or a 0. Think of a water dam: In the Off position, no water is flowing through; in the On position, water is flowing freely. Shrink this analogy down to a transistor: In the Off position, no electrons are flowing through; in the On position, electrons flow freely. (Figure 6-1 shows the *source* where the electrons come in, the *drain* where the electrons flow out, and the *gate* that switches the transistor on and off.)

A transistor is the building block to the modern processor, and boy, does it take a lot of them to add up to something useful — 125 million transistors in Intel's Pentium 4 Prescott 3.4GHz processor, to be precise. The smaller the transistor, the smaller your electron *gate* — which means faster switching between on and off, resulting in a faster overall processor. In Prescott's case, we're looking at a transistor no more than 90 nanometers (nm) across — that's pretty darn small! And pretty darn fast.

The closer you get to nano scale, the more you see evidence of a general rule: As the minimum feature size decreases, the speed and performance of micro-electronic devices rises rapidly. Computer geeks know the rule as Moore's Law.

Figure 6-1:
The side view of a transistor, showing source, gate, and drain.

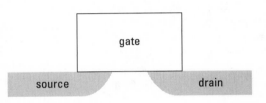

Moore's Law — as stated by Gordon E. Moore, co-founder of Intel — claims that the rate of advancement in processor technology doubles every 18 months. This dramatic notion can mean many things, but has generally come down to mean that the number of transistors squeezed onto a processor doubles every 18 months. That's been going on for about 40 years.

At the end of 2004, chips were being fabricated with dimensions in the 130nm and 90nm ranges, with the goal of 65nm at the end of 2005. There will be some problems going down to the 45nm and 30nm realm, which nanotechnology will help solve in many ways — from new mixtures of metals that make up traditional transistors to entirely new concepts, including single-electron transistors (which work like a dam that lets one drop of water through at a time — tough stuff to make real). Figure 6-2 shows a side view of Intel's Prescott transistor — an example of the former approach. Here, various metals (silicide, silicon, silicon oxide) make up a transistor whose gate is only 50nm across. A performance-enhancing feature of this transistor is the *strained* silicon — individual silicon atoms stretched apart so electrons flow through faster with little resistance — improving processor speed. Clever, eh? Chalk up one more benefit of nanotechnology — using a combination of chemistry and material science at the nano scale to produce a nano-size transistor. Now where the heck did they get the teeny-weeny pliers for the job?

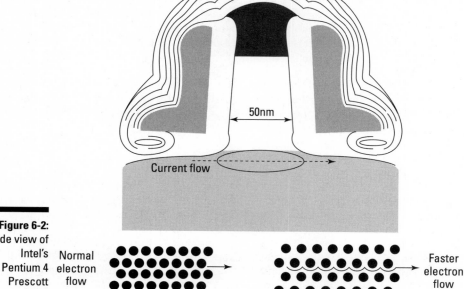

Figure 6-2:
Side view of
Intel's
Pentium 4
Prescott
transistor.

From FETs to SETs

In the world of computer processors, transistors are our ever-present on/off switches — switches regulated by tiny gates that either hold up or let pass any traveling electrons. (Refer to Figure 6-1 for a peek at what a tiny gate might look like.) To be more precise, this type of transistor is called a *field-effect transistor* (FET) and is used in much of today's electronics, including today's PC. In the push to make transistors for computer processors "better, faster, stronger," folks have come up with some unique remakes of the current transistor but have also pushed a new concept that has only recently been realized with nanotechnology: the single-electron transistor (SET). In this section, I fill you in on both approaches.

The current silicon-designed FETs are going to be hitting their physical limits — not just the limit of how small we can make them but also problems with heat dissipation, wire connections, and the materials we use to create them — within the next ten years. Therefore, scientists are exploring new ways to design and create FETs at the nano level without restricting themselves to silicon alone.

Sizing up the nano-world

Metrology, the study of measurements, is set to be a significant enabler of any advances in computer technology (well, nanotechnology, period) because metrology concerns itself with a crucial question: How do you know you're achieving your nanotechnological goals if you can't measure your progress?

To illustrate this problem, take a look at a silicon wafer. Silicon is grown into a crystalline structure — basically an incredibly flat piece of glass. How can you determine — both quickly and accurately — that your wafer is perfectly flat at the atomic level? Any imperfections at that level will propagate through your entire fabrication process, destroying your processors and cutting into your costs. (Talk about pressure!) Small wonder that the International Technology Roadmap for Semiconductors points out the need for an "inline, nondestructive microscopy process [at nanometer] resolution," meaning that any measurement you take must not destroy what you're trying to measure (and that's a risk when you have to bombard things in order to see them). What a conundrum.

Metrology-specific companies are going be the first to emerge and have viable products. Think of these companies as playing the role for nano-technological industries that America Online (and other Internet service providers) played for the Internet back in 1994. Without ISPs, no one would have been connected to the Internet and online businesses would have faltered. Similarly, without the appropriate metrological tools, nanotechnology as an industry is going to remain stuck in neutral, as businesses see no way to produce nanotech products they can neither see, measure, nor manipulate at this level.

Three companies making headway in this regard are Veeco, Nanometrics, and Zyvex. Veeco [VECO] is known for its microscopes — seeing at the nano-scale level. Nanometrics [NANO] (now *that's* a nice ticker symbol) is metrology-specific to the semiconductor industry — measuring at the nano scale. Nanometrics and August Technology [AUGT] announced their merger agreement in late January 2005. Zyvex's (private) initial goal — to make molecular assemblers — has since evolved into a quest for nanomanipulators. Well, why not? To make anything at the nano scale, you have to manipulate it. All in all, these companies will be some of the first to develop products that you may never see but will aid companies to develop products that you will buy.

One area of interest involves creating a transistor using our beloved carbon nanotube. Phaedon Avouris and his team at IBM have made significant progress along the carbon-nanotube road over the last few years, culminating in the world's smallest solid-state light emitter. "Great," you say, "now that we have a nano-light, we can actually see what's going on and get something done." Well, sort of. First, check out what Phaedon and his buddies at IBM have accomplished so far:

✔ **Constructive destruction:** Single-wall nanotubes (*SWNTs* for short) are the greatest thing since sliced bread, but even we'd admit that they do enter this world as an awful sticky mess — which makes it cumbersome to separate the metallic SWNTs from the semiconducting ones. (You'd want the semiconducting ones for any transistor. For more on how awfully sticky SWNTs are just out of the oven, check out Chapter 4.) So, what IBM did was just get all the nanotubes together, lay them on a silicon-oxide wafer, attach electrodes over the nanotubes at each end, and then "turn off" the semiconducting nanotubes — that is, they blocked any current from traveling through them. Finally, they applied an "electric shockwave" to the wafer — destroying some of the nano-tubes (the metallic ones) but leaving intact the semiconducting ones — the ones that had been "turned off," and so were impervious to the shockwave. This process can furnish a good supply of semiconducting SWNTs for our transistor, and — a nice bonus — you can apply it to mul-tiwall nanotubes. In effect, that means researchers can remove individ-ual shells from the tubes one by one — and so fabricate nanotubes with any desired electrical properties.

- ✔ **Voltage inverter:** A voltage inverter, also known as a *NOT gate,* is one of the three fundamental logic circuits — the other two being AND and OR. A 0 that goes into this inverter comes out as a 1 — and a 1 that goes in comes out as a 0. A typical voltage inverter comprises two types of tran-sistors: *n-type* (which channels electrons carrying electrical charge) and *p-type* (which uses electron-deficient regions — also known as "holes" to carry the current). Avouris's team devised a process that not only changes inherently p-type nanotubes into their n-type brethren, but also selectively converts *part* of a single nanotube from p-type to n-type. Such fine control at the nano level means we can end up with a sort of hybrid nanotube — one half is p-type and the other half n-type, creating a NOT gate — the world's first *single-molecule logic circuit.*

- ✔ **Carbon nanotube field-effect transistor (CNFETs):** These babies already outperform leading silicon transistor prototypes by achieving higher *transconductance* — the measure of how much current they can carry. (We're talking more than twice the transconductance per unit width.) Result: faster, more powerful, nano-size integrated circuits.

- ✔ **Solid-state light emitter:** After they got to the level where they could control the makeup of the nanotube, IBM scientists engineered this FET so they could inject negative electrons from the source and positive holes from the drain, as shown in Figure 6-3. (Refer back to Figure 6-1 to see an illustration of sources and drains. We're not talking the kitchen sink here.) The nanotube is attached by electrodes at both ends, sitting upon silicon oxide. Where the negatives and positives meet, they neu-tralize each other and generate light — light bright enough to see with the naked eye (if in the visible spectrum). Since this FET also has a gate, the gate operates the switch on and off. Now, the diameter of the nano-tube dictates the wavelength of the light it emits. The researchers, using a nanotube 1.4nm in diameter, generated a wavelength of 1500nm — in the infrared region of the spectrum, the wavelength widely used in opti-cal communications (say, your Internet fiber optics). So, not only are nanotubes useful in electronics, but they rock in optoelectronics as well.

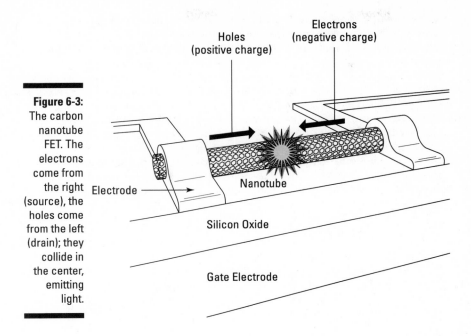

Figure 6-3:
The carbon nanotube FET. The electrons come from the right (source), the holes come from the left (drain); they collide in the center, emitting light.

Although we humans would love to use nanotubes in everything, the folks at IBM would be the first to say that those small, sticky things are a bear to work with, which limits their usefulness as a replacement for silicon in new transistors. Researchers from the Weizmann Institute of Science in Israel have addressed this issue by trying another strategy — harnessing the self-assembly abilities of (believe it or not) DNA to manipulate carbon nanotubes in FET construction. They attached DNA strands to the carbon nanotubes, hooked up their complementary strands to the electrodes, and placed a solution of the DNA-coated nanotubes on a silicon wafer. Lo and behold, the complementary DNA strands combined — and that placed the nanotubes across the pairs of electrodes.

Imagine if we, by introducing biological molecules which interacted with the DNA, could modify, in real-time, the electrical behavior of these devices. That could be very useful but would take many years to refine. These nanotube FETs show tremendous promise — and they actually *show* promise, as opposed to being a figment of someone's imagination.

In the meantime, while we're looking for ways to manipulate and mass-produce carbon-nanotube FETs, other researchers are looking for ways to avoid the *Schottky barriers* — areas of resistance to electrical conduction that occur where metal meets semiconductor. In our case, they occur at the electrode/nanotube interface, which hinders the efficiency of our FET. Researchers are determining under which circumstances the nanotube works best — and trying out metals (such as palladium) with an eye toward reducing the Schottky barriers.

Now, the whole idea of applying nano ideas to FETs was that "smaller" would mean not only "faster" but would also generate less heat (use less energy). Thinking back to our initial description of the transistor earlier in the chapter — where the source provides the "water" (okay, actually electrons) through the gate to the drain — you can think of this process as opening a floodgate. Lots of electrons pour from the source to the drain, using a lot of electricity and giving off a lot of heat — which keeps us from packing in the transistors as densely as we'd like: Too much heat destroys the processor. What if we could control how many electrons were going from the source to the drain — say, have them come down the line one at a time? Can do: Enter the single-electron transistor (SET).

The SET can best be described as a one-way bridge with tollbooths at each end. As electrons approach the first booth from the source, the single electron is allowed to pass across the bridge and exit through the second tollbooth to the drain. Now, to make this language a little more technical (though it sounds sort of like an adventure movie), the bridge is called an *island,* the tollbooths are called *Coulomb blockades,* and when the electron goes through the tollbooth, it's called *quantum tunneling.* Figure 6-4 illustrates how the electrons move through a SET in response to two factors: the *source-drain bias* — a difference between the source's energy and the drain's energy measured in electron-volts (eV) — and the gate's voltage. Now, applying a gate to the system may push the states down so more electrons can be added to the island — but in general, here's how the source-drain bias affects the movement of current:

- A small bias (as in Figure 6-4a) results in a Coulomb blockade. The electron lacks the oomph to pass through the blockade from the source to the drain.

- A large bias (as in Figure 6-4b) results in the transport of single electrons: The electron bypasses the Coulomb blockade, is added to the island, and passes on through to the drain.

Here's the real trick — temperature plays a large part in determining how well we can hope to control a single electron. Heat alone may provide the necessary energy to hop an electron from the source to the drain. As a result of this problem, most research of the SET has been restricted to low temperature environments (close to 0 Kelvin — roughly –273°C or –459°F — *very* cold). Although the SET is ideal and an interesting concept, there is a lot of work to do before it becomes feasible at room temperature.

Figure 6-4:
Raising our source-drain bias lowers our Coulomb blockade (a), allowing the single electron to pass (b) from the source to the island and on to the drain.

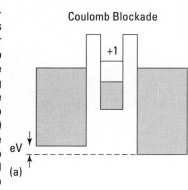

Coulomb Blockade

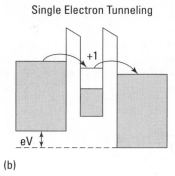

Single Electron Tunneling

Cees Dekker of Delft University of Technology in the Netherlands accomplished a room temperature SET to a certain extent. His team took a metallic single-wall nanotube (SWNT) and used two barriers to define an island between them. They created these barriers by using an AFM tip to "buckle" (selectively cave in) the nanotube. (See Figure 6-5. AFM is short for *Atomic Force Microscope.* For more on AFM tips, see Chapter 3.) The buckled areas act as nanometer-scale tunnel barriers that block electron transport, giving the island a length between 20 and 50nm. Dekker demonstrated that varying the gate voltage opens and closes the gap, which led them to conclude that varying the bias and gate voltages can give you some control over how many electrons could be let through — although not much control. In effect, the "on" wasn't much different from "off." All said, this unique concept is a good start at producing a room-temperature SET, and will undoubtedly be researched further.

Another room-temperature SET is a bit more mechanical. Researchers at Ludwig-Maximilian University in Munich, in conjunction with the University of Wisconsin-Madison, created a nano-electromechanical systems (NEMS) transistor. They created a silicon arm 200nm long and tens of nanometers across, covered the tip with gold (that's our island), and placed the tip between the source and the drain. They then applied an AC voltage to the source, matching the frequency of the current with the resonant frequency of the arm. In response, the arm started to swing between the source and drain. Result: Electrons flowed from the source, through the island, and to the drain — all at room temperature.

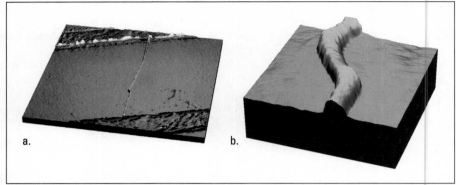

Courtesy of Cees Dekker at Delft University of Technology in the Netherlands.

Figure 6-5:
A close-up of the buckles created in the carbon nanotube with an AFM tip.

a.

b.

A more traditional route to making a SET involves a nanowire. Instead of today's top-down transistor fabrication process (see the "Fabricating new chips" section, later in the chapter), the folks at Lund University, Sweden, built their nanowires from the bottom-up, allowing the researchers to control how the different metals were introduced. This allowed them to determine the thickness of their Coulomb blockades as well as their islands. Unfortunately, these were not successful at room temperature — they needed temps at 4 Kelvin. Still, their work is controllable, impressive, and could be used if temperatures were a bit more reasonable.

Fabricating new chips

Today's computer chips (and the integrated circuits that turn up in all sorts of everyday electronic devices) have to come from somewhere; *fabrication* is the industrial-sounding word for the complex process that creates them. Your computer processor in your home computer has roughly 125 million transistors — each working in unison with the others, flipping those 1s and 0s to help you play video games or file your taxes. Each of these transistors are built in one sitting, layer by layer — basically, you have your substrate (ground layer) and you build up layers of different metals, oxides, or polysilicon, one upon the other, until eventually you get something that resembles your processor.

This isn't easy — we're talking over 250 steps over a two-week period, in an ultraclean environment, in order to produce just *one* silicon-based microprocessor. This highlights two very obvious problems in traditional chip fabrication — it takes too darn long and the working environment is a bear that just gets bigger and hungrier for resources. And, as the scale of the product gets smaller and smaller, we run into additional problems — stray signals on the chip, a need to dissipate built-up heat . . . this is getting costly, in terms of both time and money. If, as recent trends show, the smallest possible feature size decreases about 30 percent every two years — and (presumably) needs more time and money to make — then by 2015, we may hit the wall:

Making smaller, faster processors just won't justify the cost of the fabrication facility. How much is your computer going to cost when the facility that makes the chip costs $200 billion? Do any of us have that kind of money?

Before we get all Malthusian here — you know, that theory that warned that population "trends" would wreck the world's ecosystem, leading to massive starvation and the death of humankind by, say, 1923 — let's revisit today's chip fabrication processes with the added perspective of some proposed nanotechnology techniques that may be cost-effective alternatives, allowing us to continue to get smaller and faster.

Field trip to Intel!!!

Everyone off the bus — and be sure you have your partner so no one gets lost. Remember: Before we can go into the fabrication facility and into the *cleanroom* (the ultraclean environment for making the processors), we have to put on our bunny suits. No rabbit ears, fluffy tails, or hopping around — just airtight white "space" suits that protect the chips from getting gummed up with everyday human particles such as skin flakes and hair. Intel takes this very seriously, using a unique nonlinting, antistatic fabric that is worn over street clothes. The entire process just to put on the suit requires 43 steps — now we're late . . . late . . . for a very important date — in the cleanroom.

When we say clean, we really mean clean — 10,000 times cleaner than a hospital operating room. Huge air-filtration systems change the air in the cleanroom ten times every minute. Which means this guy qualifies as a "Class One" cleanroom — no more than one speck of dust per cubic foot. "Why so clean?" Pigpen asks. A speck of dust can ruin thousands of transistors. Just imagine a boulder falling on Times Square in New York, causing a huge traffic jam all across the city. The same goes for a dust particle on a microchip — it basically obstructs the chip's pathways, rendering it unusable.

Intel — and everyone else — uses one very basic technique — *photolithography* — as the basis for chip fabrication. If you've ever played with shadow puppets, you'll quickly grasp the concept. Got your flashlights ready? Here goes:

1. **Start out with a thin, silicon wafer.**

    Silicon is a semiconductor that can either conduct or insulate against electricity — ideal for our transistor's on/off states.

2. **Oxidize the silicon wafer by exposing it to extreme heat and gas.**

    The concept is similar to what happens when steel is exposed to water and forms rust. What we end up with here is a thin coating of silicon dioxide — an insulating layer.

3. **Coat the wafer with a photoresist — a substance that becomes soluble when exposed to ultraviolet (UV) light.**

UV light has short wavelengths, just beyond the violet end of the visible spectrum.

4. **Place a mask — also referred to as a stencil — of the circuit above the wafer; shine the UV light through the pattern and onto the photoresist layer.**

   The mask protects parts of the wafer from the UV light.

5. **Using a solvent, wash away the part of the photoresist that you exposed to the UV light.**

   What you get is an exposed area of silicon dioxide in the shape of the circuit.

6. **Use chemicals (acids) to etch the pattern into the silicon, removing the exposed silicon dioxide at the same time.**

   Now the pattern of the circuit is etched into the chip as lines of exposed silicon.

7. **Expose the entire photoresist to UV light, and then use a solvent to remove any trace of the photoresist.**

   You end up with a specific pattern of silicon dioxide that looks a whole lot like (surprise!) the desired circuit.

Figure 6-6 is a handy illustration of the entire process.

This entire process is repeated over and over again, with each layer of material patterned with its own unique mask and then etched away. During one of the layers, specific chemical impurities are added to the material (called *doping*) making the silicon either n-type or p-type (see the "From FETs to SETs" section, earlier in the chapter, for an explanation of n-type and p-type). Metal leads are also added to make the electrical connections.

Keep in mind that all this effort results in just one transistor — just one of the millions we need for the processor, which is just one of the hundred or so processors on a wafer. See, we told you this would be easy. Which explains why some folks might be interested in investigating some Next-Generation Lithography techniques. (Read on if you're one of those people.)

Fabrication's extreme makeover

Now, it makes sense that if we want to write smaller lines in our circuits, we should use smaller wavelengths. Current photolithography techniques (as described in the previous section) use ultraviolet wavelengths on the order of 248nm, creating line widths of 200nm. Okay, it's true that they've been able to decrease the UV wavelengths to 193nm — and create features 100nm wide — but our goal is to get down below 50nm. That's where we get extreme — extreme ultraviolet (EUV) radiation. Those wavelengths are seriously smaller — in the range of 10 to 14nm.

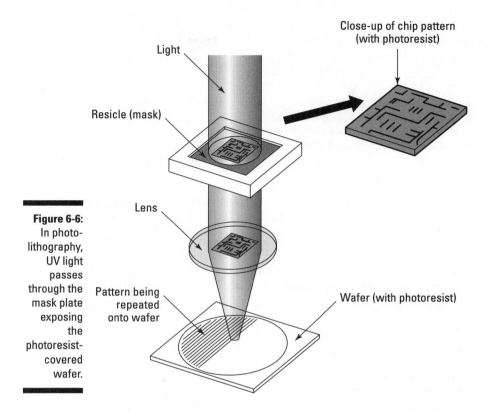

Figure 6-6:
In photo-lithography, UV light passes through the mask plate exposing the photoresist-covered wafer.

In principle, EUV lithography is quite similar to the optical projection lithography described in the previous section — it just uses shorter wavelengths. The main problem that we run into at these levels is that the properties of materials in the EUV spectrum are very different than those in the visible UV spectrum. EUV radiation is strongly absorbed in virtually all materials — which means that we have to come up with entirely new processing techniques. The following three areas have proven to be especially problematic:

✔ **Focusing optics:** Since refractive optics (such as your typical lens) absorb EUV radiation, we have to use exclusively reflective elements (curved mirrors, in other words) to focus our beam onto our sample. These mirrors require an unprecedented degree of perfection when it comes to surface figure and finish . . . which leads directly to one of our main problems with processors at the nano level — metrology. How do we know our mirrors are perfect? The reflective material doesn't interact with visible light the same way it does with EUV light; that makes measuring in the visible spectrum tricky. Looks like we have to come up with a whole new toolbox of measuring techniques for dealing with the EUV spectrum.

✔ **Masks:** The stencils used as part of the EUV lithography must also be reflective (tricky process beyond the scope of this book). This also runs into the same problem with detecting defects in the mask — how can we tweak current measuring techniques so that we can use them to work with EUV radiation?

✔ **Photoresists:** The photoresist layer is what gets exposed to the EUV radiation and washed away, allowing the underlying metal to be either etched or doped. (For more on etching and doping, see the previous section.) The thing about EUV is that it gets absorbed by all materials — which poses a problem, since we need to have a resist that is acceptable for high volume manufacture that reacts appropriately to EUV. Additionally, the photoresist must have excellent etch resistance — as we get down to the nano scale, roughness at the edges of the resist lines can create problems with exactness. We need something that can be mass-produced and can create sharp features for our transistors. (No small task.)

To address these issues arising with the use of EUV radiation, Intel has formed a consortium with AMD/Motorola, along with three national labs — Lawrence Livermore, Sandia, and Lawrence Berkeley. Considering that the EUV process is similar to lithography techniques currently in use (it'd be nice to reuse and expand upon current facilities), this may be the future fabrication technique of choice. But just to keep you on your toes, we'd like to devote the rest of this section to a few alternatives.

E-beaming

Electron-beam lithography (EBL) is a specialized technique for creating integrated circuits at the nano scale. EBL "writes" using an electron beam, in much the same way the scanning electron microscopes (SEM) from Chapter 3 "read." The SEM would bounce a beam of electrons off of a conductive sample, scanning back and forth, reading the reflected electrons as an "image." On the other hand, the electron beam (*e-beam* for short) scans back and forth writing electrons on a photoresist material, our desired circuit pattern — similar to a printer writing to paper scanning back and forth by depositing individual ink droplets. This photoresist material is now considered exposed and washed away. The previously described chemical etching process is used and our single layer is produced.

Here's what's exciting — and not so exciting — about EBL:

✔ **High resolution:** Almost to the atomic level.

✔ **Flexibility:** Can work with a variety of materials in many different patterns.

✔ **Slow:** A magnitude or two slower than traditional optical lithography.

✔ **Expensive and complicated:** The tools used can cost many millions of dollars and often require frequent service.

To give you an idea what we can do with EBL, take a look at Figure 6-7. (The folks in Professor Douglas Natelson's lab at Rice University graciously provided their talents to create this image.) Take a look at the scale on the width of the letter "D": 0.044 _m, which is 44 hundredths of a micron — 44 nanometers.

Figure 6-7:
An example of electron-beam lithography creating lines with widths as small as 44 nanometers.

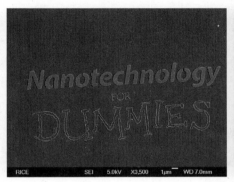 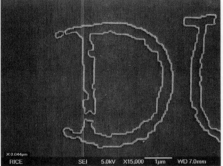

Courtesy of Sungbae Lee at Rice University.

In addition to creating incredibly small images of book titles, EBL has three niche markets:

✔ **Maskmaking:** Masks — those stencils used in traditional optical lithography — are a relative snap to make with EBL: It offers the flexibility to provide rapid turnaround time, and can control line width to stringent standards — incredibly important for future EUV and even X-ray lithography (explained in the next section).

✔ **Prototypes:** We're told that Figure 6-7 was created in an afternoon (we call that a cheap parlor trick, but okay, an impressive one). EBL aids in small volume specialty products and prototyping integrated circuits. Since EBL is flexible and provides a high resolution, it allows for quick, experimental designs.

✔ **Research:** Since EBL writes at the nano level, researchers can use it to make nano-scale structures they can use to study quantum effects and other physics phenomena. Areas of interest include ballistic electron effects (more about those in Chapter 4) and single-electron transistors — like the ones we discuss in the "From FETs to SETs" section, earlier in this chapter.

The bottom-line stuff

The business of fabrication is like . . . selling pans to the nano-gold prospectors. And, to keep this analogy rolling along just a tad longer, it just so happens that two major players in the field, Atomate and Invenios, are both based in California.

Nanotech researchers have been buying silicon-based fabrication machines and modifying them for nanotech experiments for a while — and have been spending a lot of time and money in the process. This is where Atomate comes in — they have already tweaked the fabrication machine so the researcher can actually get his or her work done efficiently (after all, time = money). Since nanotechnology is rapidly evolving, these fabrication systems require continual development and adaptation — Atomate is happy to do both. As nanotechnology moves from the lab to the factory, Atomate is poised to shift from low-cost research systems to full-scale factory fabricators. Along the way, Atomate is patenting fabrication techniques and processes, making its claim to the nanotech goldmine that much more solid.

Invenios has developed a fast, cheap process for creating 3-D shapes with a single pass of a laser. Starting with a special kind of glass whose atoms are unordered, a laser beam, guided by a computer, strikes certain areas inside the glass, displacing the atoms' electrons. When heated, the treated parts of the glass form ordered, crystalline structures which are then etched away by acid, resulting in the desired structure — from microfluidic valves (Chapter 10) to optical waveguides (Chapter 7). Thousands of components could fit on a single wafer, costing as little as 30 cents each if you buy 'em in volume.

The grab bag of chip-fabrication contenders

In no particular order, here are a few more cutting-edge techniques for chip fabrication, all vying to become The New Best Thing:

- **X-ray lithography:** Similar to the extreme ultraviolet (EUV) approach but has an even smaller wavelength (5 picometers to 10 nanometers). Its claim to fame is its penetrating power into the resist, great for creating high aspect ratios (where the height of the penetration is greater than the width). This process has been used to create molds, particularly for photonic nano-structures (Chapter 7). A couple of drawbacks to this approach include (a) the extreme cost of the X-ray source and (b) controlling the exposures.

- **Ion-beam lithography:** Similar to electron-beam lithography but instead of using electrons, we go with ions (go figure). Ions are atoms lacking one or more orbiting electrons and are already used to implant *dopants* — those impurities that make the silicon either n-type or p-type. Ion-beam lithography may offer a higher resolution than EBL (because the ions don't scatter as much as beamed electrons), but it's just as expensive and slow because of the ion sources and scanning techniques used.

✔ **Writing with AFMs:** Atomic force microscopes (AFMs) resemble a needle on a phonograph reading the atomic bumps on a sample. (Chapter 3 has more on AFMs.) The thing is, reading isn't the only thing AFMs do — they also *write*. As we discuss in the "From FETs to SETs" section, earlier in this chapter, you can use an AFM to bend a carbon nanotube in two places, creating an island. Additionally, there is dip pen nanolithography (DPN), which writes at the nano scale much the same way that a traditional fountain pen writes ink on paper. Atoms and molecules are first stored at the top of the AFM tip and then deposited on the surface of the substrate, creating the nano-scale lines and patterns.

✔ **Nano-imprint lithography:** The idea behind the printing press, Johannes Gutenberg's revolutionary invention of 1450, is pretty straightforward — mash something hard against something softer so an imprint is left behind. Stephen Chou of Princeton University has taken that lesson to heart — and has begun stamping a hard mold into a softer material — the trick here is that he's doing it with features less than 10nm across. He does this by melting the surface of a metal just long enough to press a mold of the desired features into it. His start-up, NanoOpto, has already shipped nano-imprinted optical-networking components — as well as prefab nano-channels to speed up lab-on-a-chip genetic tests. Although he's sped up his imprinting time to less than a microsecond — and can squeeze 36 times as many transistors onto a silicon wafer as is possible with today's lithography tools, this stamping process must maintain perfect alignment through the 30 stamping steps. Taking a different tack, George Whitesides from Harvard uses a rubber stamping technique: A pattern is inscribed onto a rubber surface, coated with molecular ink, and stamped onto a surface (either metal or polymer).

✔ **"Printing" circuits:** Similar to an inkjet printer, but instead of ink on paper, you print out a working ink-based plastic processor. Joseph Jacobson of MIT Media Lab's Nano Media Group has printed thermal actuators (a sensor that moves when heated), linear drive motors (small pistons), micro-electromechanical systems (MEMS — discussed in Chapter 8 and used in pacemakers, airbag deployment, sensors, and communication), and transistors. Essentially what he has come up with is a nanoparticle-based ink — made of nano-size semiconductor particles, suspended in a liquid, and sprayed onto a plastic sheet using (for the most part) an ordinary inkjet printer. The advantage of layering plastic over silicon is that the final product is physically flexible (you can bend and roll it like paper), but the disadvantage is that the processing power is slower. Future applications include computerized clothing, disposable cellphones, digital cameras the size of a credit card, and wallpaper that doubles as a television screen — all quickly printed on rolls of film.

Does a Nano-Size Elephant Ever Forget?

Only the little things . . .

Man, was that a bad joke! ("What joke?" you ask.) There is a method to our madness — it's rooted in the vital importance of memory to intelligence: What you remember has a huge influence on what you do. In an effort to explore the depths of this human question, Philip K. Dick wrote the novelette *Do Androids Dream of Electric Sheep?* — which later became the 1982 movie *Blade Runner*. The androids in this story "replicated" humans to an amazing degree, so much that they managed to get saddled with all the stuff that bums humans out — fear of death, an overwhelming sense of life's uncertainty, and lost (or untrustworthy) memories.

We posed the elephant question (an elephant never forgets) to draw a parallel: the importance digital technology is going to have when it comes to saving our own memories from being lost. Yeah, computers are going to help, but as our stock of "memories" — data — increases exponentially (we all know something about the data explosion of the last decade, right?), our need for physical memory to plug into our computers will increase just as fast. Given that fact, it should come as no surprise that some folks are looking beyond traditional memory technology — the ubiquitous random-access memory (RAM) — and that some are in fact exploiting nano-scale science in their efforts to improve upon RAM's shortcomings.

Magnetic Random-Access Memory (RAM)

Random-access memory (RAM) is a memory capability that can access data anywhere on the storage medium — hence *random* access (well, yeah — as random as my elephant joke). Imagine your data arranged in a spreadsheet table: You can access a bit in any memory cell directly if you know its address (where the row and column intersect to form the cell). It's like finding a location on a map — C-12 represents row C, column 12. (Shades of playing Battleship!) Input the address and you can go right to it, wherever it is.

Dynamic RAM (DRAM) is the most common RAM found in our computers, cellphones, and PDAs. But just because it's the most common technology in no way means that it's the best-designed option. (Was VHS really better than Betamax technology?) A major drawback of DRAM is that it requires a constant current to store a single bit of data. Every DRAM bit uses a capacitor — if the capacitor is filled with a charge, the bit is a 1; if it's empty of charge, the bit is a 0. (Think of this capacitor as a leaky bucket that must constantly be replenished or the data that the capacitor represents — more than likely a 1 — will be lost.) The capacitor charge must be refreshed thousands of times per second, and all that work eats away at the power supply.

This constant need for energy is certainly not one of DRAM's major selling points. (No one likes to admit that their product sucks the life out of portable electronics by straining a device's batteries.) Instead of constantly having to fill a leaky bucket — the DRAM model — how about making memory that only used energy when it had to *flip* a single bit? Magnetic RAM (MRAM) does just that — it uses a little bit of energy to help store massive amounts of data. Magnetic storage devices (such as your computer's hard drive) exploit a magnet's polarity — positive north or negative south — to represent a 1 or a 0. Now, as we get smaller — nano-scale small — we can capitalize on *spintronics* — the science of using the spins of individual electrons (one of the more useful idiosyncrasies of the quantum world) to help us out.

Sound intriguing? Then copy down the following bullet list and tape it to the side of your own computer monitor:

- **MRAM uses less energy:** Unlike DRAM, the only electricity that MRAM uses is to switch the polarity of each memory cell; it doesn't have to be periodically refreshed.

- **MRAM is faster:** Because MRAM uses less electricity, it cuts out the delays associated with transferring electricity between the power source and the chip — resulting in speeds 30 times faster.

- **MRAM is nonvolatile:** MRAM retains memory even when no electricity is flowing into the system — which makes it similar to Flash memory, the memory sticks you use for your digital camera. MRAM's electronic cost savings arise from the magnet's *hysteresis* — the magnetic effect doesn't disappear when an applied magnetic field is withdrawn.

Just so you know we're not talking about some pie-in-the-sky stuff, both IBM and the Naval Research Laboratory have gone ahead and developed magnetic RAM technology. In the next sections, you get a chance to examine their (differing) approaches a bit more closely.

Magnetic Tunnel Junction (MTJ)

IBM has been developing the *magnetic tunnel junction (MTJ)* since 1974, but it wasn't until 1998 that they successfully demonstrated a working MRAM. MTJs consist of two ultrathin magnetic metal layers separated by an insulator. The insulating layer is incredibly thin — so thin that it allows electrons to tunnel through — which is why it's called a *tunneling barrier*.

Figure 6-8 shows two separate instances of an individual MTJ memory cell representing one bit. The thin magnetic layers contain electrons whose spins are aligned — either all "up" or all "down." They work in one of two ways:

- **Antiparallel:** At the left of the figure, the spin of the top magnetic layer is opposite from that of the bottom — it's "antiparallel," so it resists being tunneled through. This *tunneling magnetoresistance (TMR)* occurs when the spin of the magnets are opposite; when an electric current tries to tunnel through the layers, it can't — the result is a "closed" junction.

✔ **Parallel:** At the right of the figure, the polarities of the magnet are arranged in the same direction — they're "parallel." This allows current to flow freely across the junction. Writing data to a cell consists of applying a current to the specific bit, which then generates a magnetic field that changes the spin orientation (from up to down or down to up) of one magnetic layer.

Figure 6-8:
An MTJ
memory
cell,
representing
a closed
junction
(left) and
an open
junction
(right).

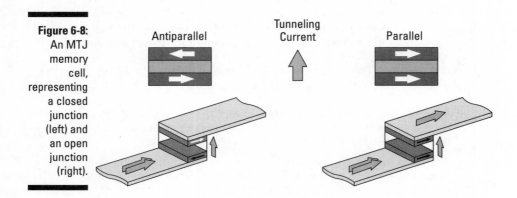

Figure 6-9 illustrates a macro (larger-scale) view of how the MTJ fits into the MRAM architecture. When the computer reads a particular bit, it determines the Word line and the Bit line — and then reads the intersection (as shown at left). When the machine writes (as shown at the right of the figure), it selects a Word line and uses the Bit line to write the polarity.

Figure 6-9:
Reading
(left) or
writing
(right)
an MTJ
memory-
cell bit in
the MRAM
architecture.

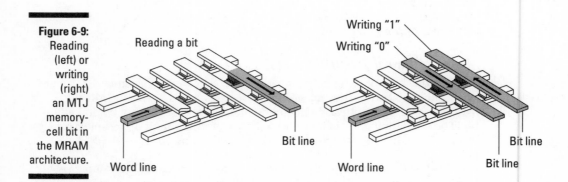

IBM's research team has taken their MTJs and run with them in several promising directions:

✔ **They've introduced lithography techniques:** Creating MTJ memory cells on a silicon substrate using photo- and electron-beam lithography has allowed not only for a mass production techniques but also quality control, creating uniformly identical MTJs across the entire working surface.

✔ **They've reduced device resistance:** Resistance of the MTJ tends to increase as the junction dimensions decrease, but we want an MTJ of a lower resistance if we ever want to realize the goal of making the fabrication of high-density chips practical. IBM has done this by controlling the thickness and quality of the tunneling barrier, reducing the resistance from 1 billion ohms (a measure of resistance) to 60 ohms per square micron.

✔ **They've achieved read and write times as fast as 10 nanoseconds:** We're talking four times faster than DRAM. This was a direct benefit of high TMR and low device resistance.

✔ **They've increased thermal stability:** *Thermal stability* (how well a chip can keep its data intact at higher temperatures) has increased from a level of 100°C to 250°C (but it will need to be higher for today's applications).

Vertical Magnetoresistive Random-Access Memory (VMRAM)

Say what? It's a whole different architecture. Vertical Magnetoresistive Random-Access Memory (VMRAM) was the brainchild of the Naval Research Laboratory and Carnegie Mellon University. The basic design and goals start off similar to IBM's MTJ:

✔ Each memory cell contains two magnetic layers that sandwich a non-magnetic conductive layer.

✔ Electrodes are connected at the top and bottom of the magnetic stack. (Refer to the stack shown in Figure 6-8.)

✔ The memory cells are formed into an array that consists of a Word line and a Bit line, as in IBM's MRAM architecture (see Figure 6-10).

This is where the similarities end. In the VMRAM design, each magnetic memory element is round and flat like a washer, which orients the magnetic moment so it flows in a circle — either clockwise or counterclockwise, depending upon whether the represented bit is a 1 or a 0. (The *magnetic moment* is a vector quantity with a direction perpendicular to the current pulse.) The left side of Figure 6-11 shows the switching of the memory element: As the current pulses in the Word lines, it generates a radial magnetic field, rotating the magnetic moment into the radial direction. This can be reversed by changing the direction of the current — reversing the radial direction. The threshold of the current magnitudes for switching the magnetic moment direction is proportional to the thickness of the memory element. (See the right side of Figure 6-11.) The thick layer on the bottom stores the bit state (either a 1 or a 0) depending upon the orientation of the magnetic moment (clockwise or counterclockwise).

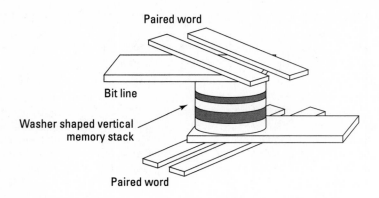

Paired word

Bit line

Washer shaped vertical
memory stack

Paired word

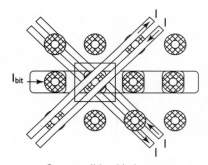

I_{bit}

One possible chip layout

Figure 6-10:
The
VMRAM
memory cell
(left) and the
VMRAM
architecture
(right).

Magnetism is confusing enough as it is, but we think you're up to taking a
look at how this scheme reads and writes (as fast as 1 bit per nanosecond —
the VMRAM's reading and writing abilities, not how quickly we can explain
said abilities!).

✔ **How VMRAM Reads:** Two electrical pulses are sent one after the other
through the Bit line. These pulses of current switch the magnetic
moment of the thin layer but not the thick layer. However, as the thin
layer switches, the current checks the orientation of the thick layer to
see which way it wants the current to go. The difference between the
first and second current pulses determines which way the current flows
in the thick layer — and that, in electromagnetic terms, is its memory
state: Clockwise is a 1 and counterclockwise is a 0.

✔ **How VMRAM Writes:** Here the approach is a lot simpler — a large jolt of
current is used to rotate the thick layer's magnetic moment. It's the thick
magnetic layer that actually stores the data. The thin magnetic layer is
used to read the data stored in the thick magnetic layer.

May the best MRAM win

There are plenty of companies including Hewlett-Packard, Samsung, Sony, Intel, IBM, Motorola, and numerous small companies (NVE Corporation and Freescale Semiconductor, a Motorola spinoff) working toward large capacity MRAM, with each one working feverishly on its own approach. (Remember, the MTJ and VMRAM models we describe in this chapter are just two approaches among many; we could easily have covered half a dozen more if we had unlimited time and page count.) Just to show how wide-open the field is, in May 2004, *Technology Review* magazine identified patent number 6,545,906 — Motorola's technique for accurately writing data to MRAMs, making it one step closer to being commercially viable — as one of the industry's "Five Killer Patents" — and we didn't even have room to cover it here. Regardless of which design "wins," we're all winners when we have fast, energy-efficient memory — and when we don't have to wait five minutes for our computers to turn on.

Figure 6-11:
Switching magnetic moments of our disks (left) and the difference in thicknesses of our magnetic washers (right).

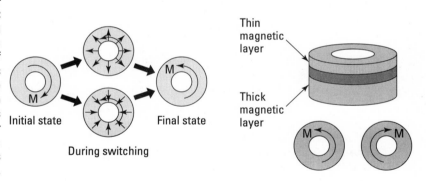

MRAM is poised to replace DRAM and become a $50 billion industry by 2010 — but the question on every investor's mind is, "Which MRAM type?" IBM's MTJ may be easier to fabricate (allowing for mass production), but the MTJ model may have problems with its linearly oriented design. This linear design may open up the possibility of a stray field (outside the memory cell) and a demagnetization field (within the memory cell) generated from the magnetic poles formed at the ends of the memory cell. This can greatly reduce its packing density and memory capacity. VMRAM is robust but may suffer from too low a resistance — no resistance means no bit-state signal, so no 1 or 0 — and no computing gets done. The MTJ may be the first to become available to the consumer — and is likeliest to be applicable to consumer applications (such as laptops). VMRAM, on the other hand, is so robust that it resists change even from hard radiation — great for military and space applications. No wonder it's being researched by the Navy.

Oh, yeah. We forgot . . .

Before we let you bet the family farm on MRAM, there are a few other approaches that you might want to consider.

Millipede drive

No, we're not talking 1980s video games or creepy-crawly bugs in tiny cars here. There really is a cool, futuristic nanotech angle here — but first, a blast from the past . . . punchcards!

Yep, punchcards — stiff cardboard cards with patterns of little holes in them — are on a comeback trail of sorts. They were the recording medium of choice during the ancient golden age of computing (from the 1920s to the 1950s). Well, nanotechnology is looking to bring back that idea in a "big" way (the capabilities will be big, anyway) — without the hanging chads. Behold: IBM has developed a memory-recording technology called "Millipede" that's loosely based on the idea of punchcards. That may sound archaic, but imagine data-storage density of a trillion bits per square inch — that's 25 million printed textbook pages on the surface size of a postage stamp. Judge by results.

The key to the idea is to make the punches nanoscopically tiny. Millipede hardware looks like a loom with a thousand tips marking 1s and 0s, as shown in Figure 6-12. It's these thousand tips that do the heavy lifting here, writing data to a polymer plate below them.

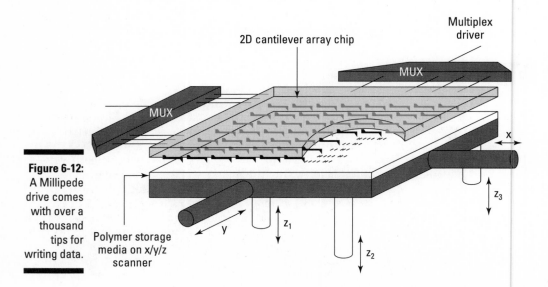

Figure 6-12: A Millipede drive comes with over a thousand tips for writing data.

IBM is excited about Millipede for the following five — count 'em, five — reasons:

- **Millipede — small in size but large in capacity:** Three billion bits of data can fit into one hole in a standard punch card. A single writing tip makes indentations only 10nm in diameter and has the potential to *address individual atoms* — which translates into a huge capacity: 200 gigabits of data per square inch, compared to what you can put on a magnetic disk drive (35 gigabits per square inch).

- **Millipede is *fast:*** Each tip supports data-transfer rates between 1 and 2 megabits per second. The thousand tips that make up the Millipede work in parallel, working smarter, not harder (just like doing laundry in several machines at once).

- **Millipede is low-power:** Only 100 milliwatts of energy, which puts it in the range of current flash-memory power consumption. Each tip is given a working surface of only 100 square micrometers — not much real estate, but those short distances ensure low power consumption.

- **Millipede is nonvolatile and re-writable:** It doesn't lose memory when the power is turned off — and each indentation can be written over.

- **Millipede is *cheap* to build:** IBM uses current fabrication technology originally designed to build computer chips. No exotic new machinery means a low manufacturing cost.

As for the details on how the whole contraption works, take a gander back at Figure 6-12 and you can see that the Millipede has two distinct layers:

The top layer contains the tips; the bottom layer consists of a thin polymer, and it moves left/right to get into position and up/down to read/write.

The tip (see Figure 6-13) is the same size and makeup as those found on the Atomic Force Microscope (AFM) we describe in Chapter 3; it writes and reads data by indenting into the polymer surface. The process goes a little like this:

- **Write:** A resistor at the end of the cantilever heats to 400°C, softening the polymer so the tip can sink into it — which generates an indentation. The dips and flat areas represent the 0s and 1s. You erase data in a similar fashion — heat up the polymer and (when the tip is removed) let the polymer fill in the former indentation like melted wax.

- **Read:** The resistor is set to 300°C (below the melting point of the plastic). This temperature does not soften the polymer, but when the tip comes in contact with an indentation, the resistor is cooled. This cooling causes a measurable change in resistance — which lets the computer know that the tip just landed in a pit.

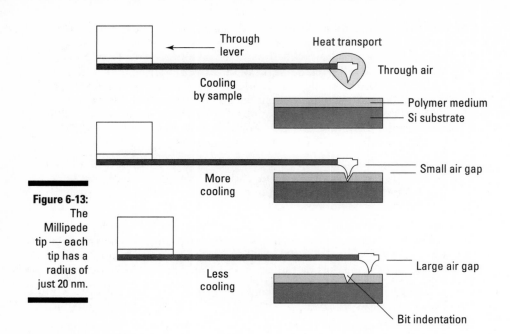

Figure 6-13:
The Millipede tip — each tip has a radius of just 20 nm.

There's one major drawback with Millipede: The tips — just like their AFM counterparts — dull over time. This will eventually make the indentation they create larger and larger, which may result in the dips overlapping, which of course corrupts the data. Just like there's a possible solution to the AFM tips problem, there's a possible solution for the Millipede problem — using carbon nanotubes as tips. Carbon nanotubes are strong and provide the nearly perfect point for tips.

Although storage is a nice perk, the Millipede has another use — can you say, "Metrology?" Metrology — the study of measures — has already made an appearance in this chapter, where a lack of a nano-appropriate way of measuring stuff reared its head. What's needed is fast, large-scale atomic imaging — and that's still one of the main hurdles facing the computer industry. What the Millipede provides is an array of a thousand AFM tips — the same tips used in atomic imaging — which could be used in large-area microscopic imaging, nano-scale lithography, or atomic and molecular manipulation. For all its small size, the Millipede packs quite a punch.

Holographic data-storage system (HDSS)

The late twentieth century had some pretty fancy dreams. For example, the 1978 movie *Superman* flew fantastic special effects at audiences faster than a speeding bullet — including a glowing green crystal (a gift from Superman's alien father) that created the Fortress of Solitude at the North Pole — complete with a data-retrieval system that housed the entire history of Superman's home planet *and* all the engineering smarts needed to create that crystal palace. (Automatically, at that.) Impressive as that bit of techno-fantasy was, part of it may actually be possible: We're willing to bet its technology would have rested in holographic storage — and we have a real shot at that. We'll only get there, however, with cost-effective components and alignment derived from nanotechnology.

As it exists today, a holographic data-storage system (HDSS) hits on some memory-storage buzzwords: more information, smaller space, and faster data-transfer times. A crystal the size of a sugar cube could contain 1 terabyte (TB) of information — that's 1000 gigabytes (GB) . . . 1 million megabytes (MB) . . . 1000 CDs . . . 200 DVDs! Not only is the capacity there, the transfer rates are blazingly fast — 1GB of information per second — that's the equivalent of transferring an entire DVD movie in 30 seconds! (Come to think of it, this century has its own fancy dreams — and some of them are already coming true.)

How does this technology do all these things? Well, it starts by adding a dimension: Instead of using a flat disc such as a traditional CD — essentially a two-dimensional medium — holographic storage ups the ante by going 3-D. You don't need to wear special glasses to view this stuff, though. Instead, the storage medium is itself three-dimensional (as detailed in the next paragraph) — and whole pages of information can be read in quick flashes of light. That's a departure from traditional ways of reading data (usually as a series of 1s and 0s such as 10011010101 . . .). Instead, holographic storage is read following a crossword-puzzle pattern (see Figure 6-14). The data is still represented in binary form, but it appears as an entire page, rather than as a linear stream of bits.

Figure 6-14:
A page read or written in holographic storage (left) with the equivalent binary representation (right).

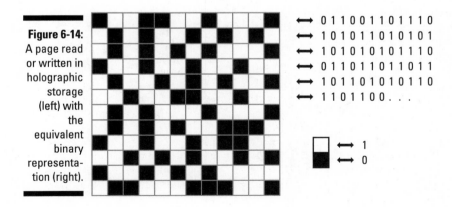

What kind of storage medium could handle the job? Well, think of a cube — a stack of pages, each layered one on top of the other, and the stack of slices resting on a photosensitive polymer cube. How do we read/write data to this guy? Figure 6-15 gives the hint of an answer, showing the general layout of an entire HDSS design. The thing to keep in mind is that there are two laser beams hitting the cube (Figure 6-15) at right angles — the *object beam* displays the entire page of data and the *reference beam* selects a layer of the cube, and that's where a page of data is either read from or written to. The meeting of the two beams is what creates the hologram.

This concept is easier to digest if we break down the components of the layout. (Okay, it'd be even easier if we could make it into a music video with a "Follow the Bouncing Ball" motif, but we haven't developed the technology yet.) Here's what we have in the Parts Is Parts Department:

- ✔ **Laser:** The originator of the light beam.

- ✔ **Beam Splitter:** Does what its name implies — splits the laser beam into two beams (the object beam and the reference beam) oriented at right angles to each other.

- ✔ **Spatial-Light Modulator:** This is actually a liquid crystal display (LCD), similar to those used as computer monitors or added to the back of digital cameras. This LCD displays the crossword puzzle layout in Figure 6-14; the laser shines through it like light through film in a movie projector. This page of binary information makes its way to the cube.

- ✔ **Scanner Assembly:** A mechanical scanner that changes the angle of the reference beam. Changing this angle allows the slices of information to be layered on the cube.

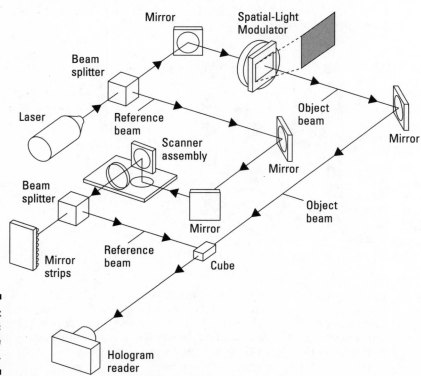

Figure 6-15: Holographic storage layout.

✔ **Cube:** Either a light-sensitive crystal or photopolymer. In either case, it reacts to the light from the laser to store the layers of information.

✔ **Hologram Reader:** A *charge-coupled device* (CCD) camera. This is actually the same technology as that used by digital cameras to capture images. The reference beam reads the data as it selects and projects the appropriate holographic layer upon the CCD reader.

After the laser projects a beam, the beam splits in two — and then heads toward the crystal from two different directions. Then the scanner assembly calibrates the reference beam and determines which layer of the crystal will hold this specific page of holographic data. The object beam makes its way through the spatial-light modulator, projecting the data onto the crystal. The data is read when the reference beam projects the crystal's holographic layer onto the hologram reader.

Okay, reality check: We're nowhere near the glowing-green-crystal level of technology yet. Two substantial hurdles are currently standing in the way of a truly viable HDSS:

✔ **Getting the alignment just right:** Aligning the laser, spatial-light modulator, and detector in a robust, cost-effective manner poses a huge problem — without proper alignment, when the laser goes to read data, other data nearby could also be read, leading to corrupt information.

✔ **Coming up with the best crystal material for a storage medium:** The crystal-cube material has yet to be narrowed down. Various photosensitive crystals and polymers have been researched but have yet to yield that magic, cost-effective mix of performance, capacity, and price.

But we're getting there. Nanotechnology will not only aid in cheaper components but greater control over the composition of the crystal material and alignment of those components — differences in nanometers can propagate through the system and throw off the entire design.

Even so, HDSS is no careless product of wild imaginations. The concept of holographic storage has been around since it was proposed in the early 1960s — but it wasn't until recently that the cost of the components dropped to reasonable levels. The laser is the same type used in common CD players, the LCDs are the sort you'd find in many higher-end consumer electronics, and the CCD sensor is what you'd find in digital cameras manufactured during the last decade. The parts are there; almost the entire HDSS can be made from off-the-shelf components.

But wait, there's more. Not only does HDSS help out big time in the Reduced Space and Faster Speed departments, but it also ratchets up our ability to pursue *data mining* — the sifting through of large amounts of information in our never-ending struggle to find even more relationships and patterns. Here's where holographic memory offers a unique advantage: *associative retrieval,* which is comparing data optically without having to retrieve it and process it. Older methods of data retrieval are *so* last year by comparison — so anything that advances data mining is set to make a big splash.

No, Lex Luthor, Incorporated, won't be making that splash (Superman can relax). Three real-world companies hope to put out holographic storage by 2006: InPhase Technologies (a Lucent spinoff), Aprilis (a Polaroid spinoff), and Optware (a Japanese company). Their initial target market includes those companies and government agencies that do image-search-and-comparison work for the defense and intelligence communities.

At least one company is taking a run at one of the big HDSS hurdles: the photo-sensitive polymer to be used for the crystal. In 2002, Aprilis discovered a silicon-based polymer that holds three to four times more data than other current holographic media. What's more, the new material doesn't shrink as much when struck by the laser (good thing, too — shrinkage introduces error). Good news, to be sure — but even with this and other advances, the photosensitive material is not yet rewritable. But you can bet they're working on it.

DVDs

DVD manufacturers are not just sitting on their hands, waiting for the newer technologies to come in and eat their lunch. They are busy themselves, trying to come up with their own New Best Thing. Current DVD technology uses red lasers (red wavelength is about 650nm), which limits their capacity to about 4.7GB per layer — and that translates to about eight total hours of video. Lowering the wavelength, however, can do wonders for increasing capacity. A blue laser's wavelength is on the order of 475nm — and shorter wavelength equals greater capacity. How much greater? Well, when you look at the two formats that take advantage of this technology (Blu-ray and HD-DVD), you can see quite a jump: Blu-ray is already getting 27GB per layer, and HD-DVD is up to 20GB per layer.

Can they really do that?

Just for the sake of argument, imagine taking the blue laser a step further. Imagine researchers at the Georgia Institute of Technology creating nanoclusters of two to eight silver atoms apiece by shining a blue laser beam on an ultrathin film of silver oxide. Each of these nanoclusters could then serve as a single bit — the smaller the bit, the larger the storage capacity, allowing for *thousands of gigabytes* of data. Imagine that data could be read by exposing the clusters to green light, causing them to glow. Additionally, these clusters could be made to glow in different colors, opening the possibility to simultaneously store multiple bits in the same cluster. (Actually, you don't have to overwork your imagination here; the Rambling Wreck from Georgia Tech has created all this stuff already.)

Phase-change memory

Phase-change RAM (PRAM) stores data at the nano level by using a change in the phase of the material to represent a binary digit. The material used here (called chalcogenide) is the same stuff currently used in rewritable DVDs. Figure 6-16 shows the two phases that are possible for the chalcogenide material — it can be as straight as uncooked spaghetti or as intertwined as a pile of noodles. The crystalline phase is ordered, the amorphous phase is disordered — and they're as different as (hey!) a zero and a one.

So, how do we get from stuff that looks like spaghetti to something that can write and erase data? We're glad you asked:

✔ **Getting "spaghetti" to write data:** Tiny amounts of chalcogenide are heated (with either a laser or electricity) to a relatively high temperature. It is then quickly cooled, freezing the material in the amorphous state (spaghetti, intertwined lines).

Figure 6-16:
A polymer matrix showing crystalline regions (straight, parallel lines) and amorphous regions (spaghetti, intertwined lines).

✔ **Getting "spaghetti" to erase data:** The same area of the chalcogenide material is heated to a medium temperature and slowly cooled. The existing surrounding crystalline lattice provides the template from which the material re-crystallizes from the outside inward, ensuring alignment of all the lattice planes (straight, parallel lines).

Now, you may wonder how we read our bits when our spaghetti is in either a crystalline or amorphous phase. Easy. The crystalline phase reflects light and has a low resistance. Conversely, the amorphous phase absorbs light and has a high resistance. The reading may be done with a laser — or by measuring electrical resistance (albeit at temperatures lower than our write/erase temperatures). So what we have here is two ways to utilize the chalcogenide material: optics or resistance.

There are a few companies working with phase-change memory. Hewlett-Packard Laboratories are tackling the optical side. Now, as you've probably picked up by now if you've been reading this chapter from front to back, storage density is limited by laser wavelength (a few hundred nanometers). HP's approach is to use an electron beam instead of a laser — which has enabled them to come up with bits 150 nm across, providing storage density of 29 gigabits per square inch — about three times the density of current DVDs. Researchers believe they can increase this to 100 times the density of current DVDs. Samsung Electronics is taking the lead with the resistance model for PRAM, bypassing both Intel and SGS, producing 64MB chip samples. Samsung hopes to enter full-scale production by 2006.

PRAM does great when it comes to things like high speed and nonvolatility. It currently faces problems, however — three in particular: long-term stability, how many read/write cycles it can handle, and its sensitivity to high temperatures. HP is addressing the problem of read/write cycles by introducing new materials; Samsung claims its chip can store data for more than two years, is more durable than flash memory, and operates at high temperatures of 85°C.

Nanotube RAM

Yep, RAM in a nanotube. (Sounds like a 1950s tune.) Nantero, of Woburn, Massachusetts, is using carbon nanotubes as miniature mechanical relays in their memory cells. Carbon nanotubes could serve this function very well, given their resistance to temperature, magnetism, and radiation. Imagine a grouping of nanotubes acting as an actual bridge across a 13-nm-deep, 130-nm-wide canyon. Applying an electric field to the nanotube would cause it to flex downward into this canyon, where it would come in contact with the floor — in this case a metal electrode. The van der Waals' forces that make the nanotubes inherently sticky — even with each other — would hold the switch in place (nonvolatile memory! yay!) until a field of different polarity caused the nanotube to straighten out. Thomas Rueckes (a former Harvard graduate student for Charles Lieber) and his team at Nantero have ironed out the fabrication process, but they still need to increase their large-scale reproducibility; they hope to have nanotube RAM ready for commercial production by 2006.

Molecular memory

Hewlett-Packard Laboratories (Philip Kuekes and Stanley Williams), along with the University of California at Los Angeles (James Heath), decided to throw out silicon and get really small: They started working with molecules. This molecular memory device is arranged with a crossbar array of nanowires sandwiching molecules that act as the on/off switches, as shown in Figure 6-17. At each crossroad intersection in the array, you'll find a molecule with a particular electrical resistance. A specific voltage is used to decrease this electrical resistance so the molecule can represent a 1; another specific voltage increases the resistance so the molecule can represent a 0. This concept is just the tip of the iceberg for molecular electronics — and our next chapter has an entire section devoted to it.

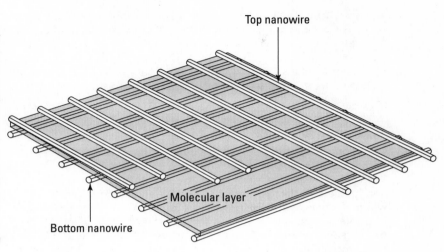

Top nanowire

Molecular layer

Bottom nanowire

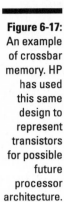

Figure 6-17: An example of crossbar memory. HP has used this same design to represent transistors for possible future processor architecture.

Quantum Leaping (Oh, Boy . . .)

Ever since physics wandered into the weird world of quantum mechanics about a hundred years ago, science fiction has loved that weirdness and borrowed its language. Case in point: *Quantum Leap,* a TV show aired from 1989 to 1993. The main character, Dr. Sam Beckett, leaps through time, landing (due to some quantum weirdness or other) in other people's bodies so as to "fix" history. He is accompanied by Al — the hologram of a friend from the future (who only appears to Sam) — and Ziggy, Al's computer (dare we say "quantum computer"?). Ziggy determines what's wrong and what Sam has to fix. Sam always has a dual identity: Inwardly he stays himself; outwardly he looks like (for example) a baseball player, a test pilot, even Lee Harvey Oswald. He's Sam and not-Sam at the same time. (Hold that thought.) We're going to use Sam, Al, and Ziggy to explain some quantum mechanics — oh, boy.

Quantum. Practice saying it with a straight face ("*quant*-um") at your next cocktail party. (You'll feel smarter. You'll look smarter. You'll change the subject a lot.) Quantum mechanics (or *quantum physics* or *quantum theory*) describes the physical interaction between atoms more accurately than classical physics. Okay, the apple-bonking-Newton-on-the-head stuff works fine for the everyday world — but for the nano-world, a few weird notions are the order of the day:

✔ A quantum is a discrete unit in the same way an individual electron is the basic unit of electricity or an individual photon is the basic unit of light; if you have more than one present (and you usually do), the plural form is *quanta.*

✔ Atoms vibrate and interact with other atoms, creating a whole slew of unique electrical and magnetic properties.

> ✔ Electrons show properties of both waves *and* particles — which is what first put the kibosh on Newtonian physics at the turn of the twentieth century.

Okay, but what's all this weirdness good for? Well, for openers, it's vital to two concepts that we'll wrestle with later on: superposition and entanglement.

Superposition — two bits in one?

Common sense says it shouldn't happen, but it does: *Superposition* is one object simultaneously possessing two or more values for a specified quantity. When Sam — our *Quantum Leap* guy — is in a person, he is both that person and himself. Rather strange. Let's get quantized and look at what happens when electrons get their properties superposed. An electron can have spins multiple ways but we'll look at the case where it spins either up or down, a 1 or a 0 (see the connection with computers?). Now, we don't know what spin the electron has until we measure it — and the real kicker here is that when we measure it, we change its spin. This is a Catch-22 — we want to use an electron to represent a computer bit, but determining whether it's a 1 or a 0 *changes* it from a 1 or a 0 to whatever it wasn't — so we still don't know what it is. (Got an extra aspirin?) So before we try to measure the electron, we assume it behaves *a little like a 1* and *a little like a 0* — that it has properties of both a 1 and a 0 at the same time. This is called a *superposition of states 1 and 0.* Since we're talking a lot about something that can be a little bit 1 and a little bit 0 at the same time, utilizing individual electrons and photons, let's give it a name — how about *qubit* (short for *qu*antum *bit*)? Figure 6-18 illustrates the critter.

We can certainly use superposition to our advantage — for example, *spintronics* ("spin-based electronics") exploits not only an electron's charge but also its spin. Current technology requires an electron to represent either a 1 or a 0, requiring eight bits to represent numbers between 0 and 255 but only one number at a time. Ah, but spintronics let us use qubits — so, we can use the "spin up" and the "spin down" states as superpositions of 1 or 0 — in a typically weird way: Because every qubit is a little bit 1 and a little bit 0, we can use eight qubits to *represent every number between 0 and 255 simultaneously.* Yes, we can get away with that — and it allows the compression of massive amounts of data into a small area — one trillion bits per square inch. This technology will dramatically improve memory elements, logic elements, and spin transistors, possibly evolving into the much-longed-for "computer on a chip" — and eventually quantum computers. (Many of us will still need that extra aspirin, though.)

Such possibilities may raise the question of just *how* we can use the spin in the superposition state. Light of specific frequencies, pulses, or polarities can tweak the state. Since we haven't "measured" the spin states, the qubits act as if it has both values — executing two operations at once — giving us parallel computing without taking up any more space. This is some cool stuff.

Computing with Qubits

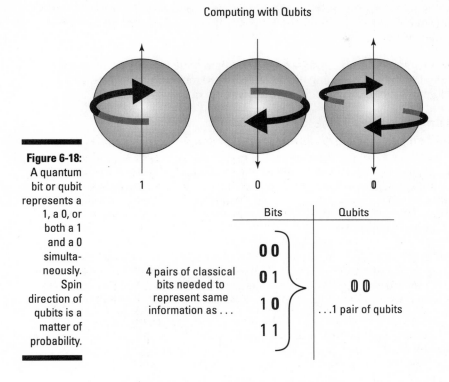

Figure 6-18:
A quantum bit or qubit represents a 1, a 0, or both a 1 and a 0 simultaneously. Spin direction of qubits is a matter of probability.

1   0   0

Bits	Qubits

4 pairs of classical bits needed to represent same information as ...

0 0
0 1
1 0
1 1

0 0

...1 pair of qubits

Entanglement — I am you as you are me

If you think superposition is crazy, *entanglement* is just bonkers: The quantum states of two or more objects are always described with reference to each other, even if they're physically separated. (Einstein actually called this phenomenon "spooky action at a distance.") If two electrons are "entangled," what happens to one *here* also happens to the other one *over there*. Here's where *Quantum Leap* comes in handy again. . . .

Suppose Sam has just finished his mission, a bright blue light engulfs him, and there he goes, "leaping" between times but still tied at both places. The information that makes him Sam is what moves — nothing else. Sam's *atoms* do not leap; the *information* — the quantum states of each atom and electron in his mind — moves because it's entangled at either end.

Okay, okay, we know time travel is fictional. But what actually happens when two electrons are entangled is about that weird. The information moves, but you still need two different particles.

Why two? Well, another science-fiction device offers an illustration: *Star Trek's* teleportation technology. Suppose that famous starship arrives at the planet Quantum Entanglement. A luckless, red-shirted ensign, unaware that he's one of the expendable characters, steps into the transporter — but quantum entanglement works only for information, not atoms. Poof. He does not materialize down on the planet. Nobody else volunteers.

This is tricky to understand — even scientists have trouble understanding and explaining what's going on. We're going to move on to some of the applications of quantum mechanics: quantum cryptography and quantum computing.

Outwitting Alia with quantum cryptography

Suppose Sam and Al, our friends from *Quantum Leap,* are on a mission and up pops Alia, an evil leaper who sets wrong what was once right. In order to communicate without letting Alia know what they are saying, Sam and Al use quantum cryptography. To see how this might work, you should know that time-tested cryptographic systems involve using one-time pads. Here's how the one-time pads work:

- Two people have the identical keypads, each consisting of a random sequence of letters that translate into corresponding numbers (A = 1; B = 2; and so on). If the key sequence is random, so is our ciphertext, which won't allow someone to determine a pattern and decipher it.

- Encryption involves adding the message to the key and, if the resulting number is greater than 26, then 26 is subtracted from the resulting number. What we end up with is the ciphertext. In our example, X is the letter from the key and H is the letter from our message:

   X (23) + H (7) = 30; Since 30 is greater than 26, we subtract 26 resulting in 4 (E) which is our ciphered letter.

- To decrypt, subtract the ciphertext from the key — and, if the result is negative, add 26, like this:

   E (4) – X (23) = -19. Add 26 and we get 7 (H), our original message letter.

This seems incredibly simple — decoder-ring simple — especially since this method is theoretically unbreakable. However, there are four criteria for successfully using a one-time pad:

- Each party must have the same key.

- Each keypad must be as long or longer as the message.

- The key may not be reused.

- The key must be destroyed.

For Sam and Al, the most important point is the first one — a secure transmission of this one-time pad. Here's where we run into Heisenberg — or at least his famous "uncertainty principle" (which states that the measurement of one property in a quantum state will perturb another property): We can't measure two properties at the same time. This sounds a lot like our superposition — photons can be polarized either horizontally or vertically, but the act of measuring changes the whole setup. Although this sounds bad, we can cleverly use this to our advantage. Figure 6-19 shows how we can set the spins of our photons — and set the bit value each spin represents. Notice we have two ways to represent our 1s and 0s — this turns out to be important.

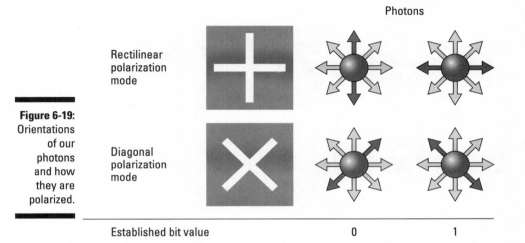

Figure 6-19: Orientations of our photons and how they are polarized.

Al has now figured what Sam must do to leap correctly and needs to relay that message to him covertly. To send the message, Al uses a photon gun (laser) and a polarizing filter; Sam and Alia, trying to intercept the message, each have a detection filter. Take a look at Figure 6-20: It shows Al, Sam, and Alia, along with their various pieces of equipment (we'll assume that both Sam and Alia have access to this advanced equipment in the time they're in). Let's take their transmission process step by step:

- Al creates a random sequence of 1s and 0s on his computer Ziggy, and translates them to the rectilinear and diagonal polarizations for each photon that the laser kicks out. He makes note of what that random sequence is.

- Sam randomly chooses which detection filter (either rectilinear or diagonal) is used to read each photon, and makes note of that as well. Then quantum mechanics come into play: When a diagonally polarized photon encounters a rectilinear detector, the photon will appear to have either a vertical or a horizontal polarization. But measuring the polarization affects the photon.

- After all the photons have reached Sam, Al shows up as a hologram and they talk to each other in the open (Alia may be spying but that doesn't matter). Sam tells Al which sequence of filters he used to read the incoming photons — but he keeps secret what the actual bit values he read were.

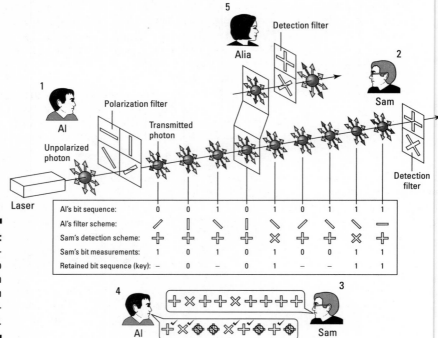

Al's bit sequence:	0	0	1	0	1	0	1	1	1
Al's filter scheme:	╱	│	╲	│	╲	╱	╲	╲	─
Sam's detection scheme:	✚	✚	✚	✚	✖	✚	✚	✖	✚
Sam's bit measurements:	1	0	1	0	1	0	0	1	1
Retained bit sequence (key):	–	0	–	0	1	–	–	1	1

Figure 6-20:
A step-by-step walkthrough of quantum cryptography.

✔ Al then tells Sam which of the filters that he used were correct. From that information, they know to disregard the filters that were not correct — and only use the correct filters. Now, both Al and Sam have a series of bit values that represent their shared key. This series of bits is shorter than the original number of photons used — but it has to be greater than the size of the message meant to be encrypted.

✔ Alia, in an effort to read the photons' polarization, also used a random detection filter and tried to resend the same polarizations she read to Sam. Let's just say that at this point, things go "a little ka-ka." Thanks to Heisenberg, since Alia measured the polarizations and picked the wrong filter, she could introduce errors — errors that Al and Sam could pick up on and disregard as incorrect. Another nice thing is that Alia can't read one photon through both filters at once — that would be measuring two properties at once, and quantum mechanics won't let that happen.

Besides outwitting Alia, the main motivation behind quantum cryptography is actually the quantum computer. When quantum computers come around, our current method of software cryptography will be moot. Quantum computers specialize in a few areas — and one of those is the factoring of seriously large numbers. Current software encryption (that is, 1024-bit encryption — the gold lock that you see on your Web browser when you pay for something online) is only secure if we have a large number that is difficult to factor.

Developing a secure cryptographic scheme is important even today — what if someone stored an encrypted message today that could be broken by a quantum computer tomorrow? Not all messages are timely — raw census data, commands for a commercial satellite, or even the formula for Coca-Cola could be decrypted at a later time.

There are some problems with quantum cryptography. Creating a device that emits _a single photon at a time_ is a big snag — but one that we're sure nanotechnology will solve. For example, we can introduce a single atom into a block of a different element — exciting that single atom will produce our single photon.

Why is a single photon important? If two photons are transmitted at the same time, Alia may use this to her advantage at eavesdropping — reading both photons with both detectors will increase her chance of knowing what the photon was and forwarding it correctly on to Sam without tipping him off that she's spying. To solve this sort of problem, scientists blend the security algorithms of information theory with the quirks of quantum mechanics, resulting in the emerging field of _quantum information science._

Two additional problems crop up, however: slow transmission rates and short distances. Since we're sending one photon at a time, science has not yet caught up with making this a quick process. Additionally, when we measure a photon (which is either rectilinear or diagonal) with our detector, we may pick the wrong orientation detector (either rectilinear or diagonal) and read the wrong position of the photon. This prevents us from resending the correct orientation of the original photon. This is similar to why we can "outwit" Alia — she can't read our photons because the act of reading and resending may screw up the original photon's orientation. Ultimately, we can't amplify our signal over great distances as we typically do with electric signals. Current electrical repeaters are scattered throughout the Internet, reading information and retransmitting them to make sure that distances don't degrade the signal.

We can send photon information two ways — through fiber optics or through air. NEC has recently pushed the limits of fiber-optic photon transmission to 150 kilometers. Los Alamos National Laboratory created a 10-kilometer link over air in 2002. So far, the fiber-optic approach has proven the most reliable.

We know what you're thinking: "Where can I get one and for how much?" For a mere $70,000 to $100,000, you too can own MagiQ's quantum cryptographic system. Of course, these won't be something you can just plug-and-play on your home computer — and at this point, we don't recommend going out and buying one unless you're a government agency or federal bank. Companies may also need this technology to do remote backups of vital business data — a safeguard that prevents putting all their eggs in one basket. MagiQ (based in New York City) and id Quantique (based in Geneva) both have commercial products in use but will upgrade to more reliable service (especially in terms of distance issues) as nanotechnology develops. Additional cryptographic systems have been developed by Japan's NEC (150-kilometer optical-fiber) and England's Qineti! (10-kilometer range through air).

The year 2014 may be when this technology hits mainstream — but quantum computers themselves won't be out until after 2025. Still, we've studied the tea leaves long enough to come up with the broad outlines of what such a computer will look like. The next section offers up to you, dear reader, our best guesstimate.

Building Ziggy, a quantum computer

Our quantum computer's architecture stems from our use of qubits and takes advantage of properties such as entanglement and superposition (business as usual in the world of quantum mechanics). Each qubit will be entangled with other qubits, and will represent a 1, a 0, *or* both simultaneously. This sounds like a sleight-of-hand explanation mainly because the traditional way to imagine a classical computer calls for transistors linked with wires, each one allowed to represent a 1 or a 0, not both at the same time. Egad, what's this world coming to? Answer: entanglement and superposition. How they apply to quantum computing is a tale in itself.

Multiple *entangled* qubits, each performing the same computation, work in parallel — just like splitting up your laundry over multiple machines. This greatly increases computing power for certain applications. Factorization — breaking down large numbers into their factors — is one such application and is our motivation for the quantum cryptography. For example, 15 can be factored into its primes (3 and 5) by almost any third grader, but factoring larger numbers can be incredibly difficult. (The number 15 may seem trivial but it's the smallest number that can be factored into two separate prime numbers.)

"What about 6?" you might ask, and we applaud your thinking. The number 6 is the smallest number that can be factored by two prime numbers, 2 and 3. All prime numbers, except for 2, are odd. So, using 6 in the following experiment is kinda cheating and not fully testing the quantum computer or the algorithm that the computer used.

Current Internet encryption is based upon a 1024-bit algorithm that uses about 300 decimal digits — and would take roughly 3,000,000 years to break into its prime factors, given today's computational technology. Quantum computers could knock that figure down to a few decades: maybe years; maybe months; maybe a matter of a few seconds. Regardless, it's turning the impossible into the possible. Other applications that take advantage of *quantum parallelism* are searching through large databases developing relationships and highly detailed simulations — and we're not talking about video games. Imagine simulating atomic and molecular interactions, giving us great detail and timely information about physics, chemistry, material science, biology, and medicine. Besides breaking codes and simulating the earth, we might see it in casinos — quantum computers are great at generating truly random numbers — something a classical computer can never do.

Quantum computations are *probabilistic* — based on the probability of whether we have a 1, a 0, or both. This is superposition of a qubit. Remember, though, that we destroy the parallel function of the qubits when we measure — instead of being both a 1 and a 0, it is defined as a 1 or a 0, but not both. So, in order to execute an algorithm:

1. **You tweak the qubits to represent either the starting 1 or 0, but they are not defined (there is a high probability that the qubit represents the 1 or 0 you want to start with).**

2. **You execute the program in parallel and rapid succession.**

3. **Finally, you *measure* the answer and a defined result is given.**

4. **This entire process is repeated until it is determined that the answer that is given represents the highest probability that the answer is correct.**

So, quantum computing is sleight-of-hand — we run the program and determine with high probability that the answer that we have is correct and what we want.

In 2001 (the year, not the movie), IBM's Almaden Research Center factored 15 into 3 and 5 using a 7-qubit quantum computer. This quantum computer utilized Shor's Algorithm, named after Peter Shor of AT&T's Bell Laboratories who devised the algorithm to factor large numbers into their subsequent primes. In essence, it followed the above concept of determining the answer with a high probability.

This not only shows that it can be done but, theoretically, is scalable using more qubits — more complex numbers can be factored into their primes. It will take several dozen qubits to solve real-world problems. A 30-qubit quantum computer would equal the processing power of a conventional computer that could run at 10 *teraflops* (trillions of *floating-point operations* per second) — the world's fastest computer, IBM's BlueGene/L runs at 70.72 teraflops. This incredible speed comes about because individual atoms can change energy states very quickly. Fast but not as fast as your brain — an estimated 10 quadrillion (one followed by 15 zeros) instructions per second.

Even with IBM's great work, we run into three problems that will slow down our advance:

✔ **The need for new algorithms:** Like Shor's algorithm for factorization, new algorithms will have to be devised for specific applications.

✔ **The need for new measuring techniques:** We'll need to discover new ways to tweak and measure our qubits. IBM, in the work above, used nuclear magnetic resonance (NMR) to detect variations of the qubits. NMR is an instrument similar to those used in hospitals for magnetic resonance imaging (MRI). As we progress toward a quantum computer, and toward advances in nanotechnology in general, you'll continue to see a crossover of technology — in this case, medicine and computing.

> ✔ **The need to counteract decoherence:** This is caused by the breakdown of the quantum properties (superposition and entanglement) as a result of the quantum system interacting with its environment. (We're told this is a very big problem.) This explains why we tend to not observe quantum behavior in macroscopic objects. Decoherence can occur very quickly while we run our quantum computer and can introduce errors into our results; by slowing down the process with low temperatures and error correction, we can minimize the effect.

That said, it's important to keep in mind that although a quantum computer will be able to perform any task that a classical computer can, it may not be very useful for every operation out there — word processing, for example. It's safe to say you probably won't have a quantum computer on your desktop anytime soon — a practical working model for the masses may not even be around for another 20 years. (Oh well. Call it something to look forward to.)

Chapter 7

Routing Information at the Speed of Light

In This Chapter

▶ Using nano-designed crystals to manipulate light

▶ Speeding up telecommunications by replacing electronics with nano-optics

▶ Revolutionizing imaging with the help of nanotechnology

*E*instein tells us nothing moves faster than light. In fact, it is so fast that as you approach the speed of light, time itself starts to slow down. (Now, *that's* fast.) So when you want your e-mail or your fax to get to somebody as soon as possible, what you really want to say is "get it there at the speed of light."

Of course, visible light is just one narrow segment of the electromagnetic spectrum — and all those waves travel at light speed. Telecommunications utilize some of these various forms of light (radio waves or microwaves, for example) to throw information from one spot on the earth to another. With billions of pieces of information flying over our heads every split second or passing below us in an optical cable, the whole world seems to be full of our thoughts. So full, in fact, that it takes a monstrous network of electric devices, cables, and computers to keep it all sorted.

And sorting takes time. Even with the best electronics today, sorting digital signals is not instantaneous, and every device, cable, or computer slows down the flow, even as it's doing its appointed task of straightening all the signals out. Where then, is that perfect device, the one that does all the work without taking any time at all to *do* said work? As you might have guessed, nanotechnologists are leading the way here, exploring ways to create an all-optical router that could route the information without having to convert it from light to electricity — in the process increasing its speed by a factor of 100! They are also looking at other applications for nano-designed optics, including ultrasharp image recorders and projectors — even a nano-size laser. In the world of nanotechnology, the future won't just be smaller; it'll get a whole lot faster!

Manipulating Light with Crystals

Simplicity is scarce and sought-after these days — but do you have to sacrifice it if you want speed? In the ancient world, all roads led to Rome, which made maps pretty simple — you could get your goods, news, and gossip from here to there in pretty good time, on roads that Roman technology made straight, firm, and easy to build. As our technology still seeks to increase the speed at which information travels, the scale gets global and we find the simple Roman roads have been replaced by the (frighteningly complex) information superhighway.

You may have heard of this "highway" and thought, "where the heck is this thing?" or "Can I drive my car on it?" The answers are: "Right under your nose," and "No, please stop trying." Although the information superhighway has often been used as just another name for the Internet, it also describes the vast network of (usually underground) optical and electrical cables now used to carry information.

Improving those cables, and increasing the speed limit on the highway, is a good way to get information traveling faster, and significant work has already been done in that arena. *Fiber optics,* using flexible cables of glass to carry information by means of light pulses, goes a long way towards moving news faster — remember: light travels faster than electricity — and optical cables can carry a lot more information than electrical ones — thousands of times more. A single fiber, in principle, can transmit up to 25 trillion bits per second — that's all the telephone conversations taking place in the United States at a given instant.

That's all fine and dandy, but we shouldn't be resting on our laurels here. Nanotechnology is set to take the *next* step and improve the highway yet again. And there is definitely still room for improvement. As you might imagine, keeping the highway from jamming when there's so much information traffic careening about is a hard job, and electrical routing is presently working at near capacity. ("Electrical *what?*" you say. Oh yeah. We forgot one little detail: Electrical routers have to convert the light pulses from the fast fiber-optic cables into electrical signals, and then reconvert and send them out into optical cables again after they've been sorted. When the routers start reaching the upper limits of their capabilities, you start getting electronic bottlenecks at the routers.)

This overcomplicated, maxed-out scene could use some simplifying. Well, how about a little crystal clarity? Crystals designed on the nano scale could replace electrical routers — and cut through the conversion folderol — by *directing the light itself* instead of first converting it into electrical signals. The fiber-optic cables you use to carry information are potentially capable of transferring data at 10 to 40 Gbps (gigabits per second), but most electrical routing occurs at less than 1 percent of that rate. If you transfer to an all-optical router, you could route most data packets in less than one trillionth of a second, pushing routing speed till it can handle the full capacity of the fiber-optic cable network. Not only will the highway be able to handle higher speeds, it will be able to handle heavier traffic. (If only the automotive highways were improving so well.)

Before we can look at the details of how such an all-optical router would work — or even the myriad other possibilities that nanotechnology can provide for the optical world — we need to look at the nano-scale pieces of light that our crystals will be dealing with. These pieces are called *photons* — which means that the science of manipulating such pieces of light is known as *photonics*.

Getting hooked on photonics

Okay, so what's a photon, and why do all the "f" words in science start with "ph?" A photon is the smallest unit of light, just like a single electron is the smallest unit of electricity. You could never see a single photon with your naked eye because it doesn't really have a shape or size, and you could never feel its weight in your hand because photons don't have any mass. They are just like light; in fact, they are light — the building blocks of light, and as such they normally travel in big groups so that you can see them all together as, say, a sunbeam.

Miss Sunbeam versus Photon Man

The information superhighway can't live on sunbeams (poetic as that would be) because such photons are usually disorganized, and who wants a garbled message? Orderly photons have to be made — and come from special sources such as lasers — before they can be used as signal carriers. The trick is, photons can't really carry anything (they'd need weight to do that) — so they aren't just the messenger, they're also the message. By varying the number of photons you send (and when you send them), you can form a code of high and low pulses, much like the dot-and-dash Morse code used in Boy Scout telegraphs the world over. Now, in the nano-crystal routers we're set to build, we'll be able to send those pulses in the right places *as light,* without first having to convert them to electricity and then back to light again.

When you work at the nano scale, you rarely encounter large mobs of photons (like the sunbeam); instead, you have to deal with a few photons at a time. If your nano-crystal accidentally stops just one of the photons, you've immediately lost most (if not all) of your information. Stopping light — as anyone who ever made shadow puppets knows — is ridiculously easy. Photons love to be absorbed by just about anything. And those things that don't absorb light usually reflect it. Thus the intrepid photon messenger must face getting either sucked up or bounced.

Controlling where photons are absorbed and where they are deflected is the business of photonics — and the concern of all nano-scale optical devices. Most of the time, we want those photons deflected as a means of moving information along a path, bouncing photons to their final destination. So we do something that bars and nightclubs started a long time ago: We check IDs.

Wavelengths: Creating nano-size IDs

To photons, the world is always vibrating because *they* are always vibrating. This shaking, rattling, and rolling is a fundamental part of any photon's life — so much so that a photon (and light itself) can be considered as a wave. Okay, how can a photon be a particle *and* a wave? Well, maybe that's the only way it can deal with constant change — by hedging its bets. After all, you — along with all other things in the universe — are made up of stuff that's both a particle and a wave. Whether the stuff *acts* like a particle or a wave depends on the situation.

Whoa — stop the ride. You came here to read about nanotechnology, not have the universe start vibrating all around you. Well, not to worry: You can safely forget about most of the universe vibrating because its vibrations are incredibly fast and really small. On the other hand, photons — traveling as they do at the speed of light — have quite detectable vibrations. In fact, the length of the space it takes for them to go through one cycle of a vibration is what determines the wavelength of a distinct bunch of photons. (So *that's* where that term comes from!)

A typical wavelength of visible light is about 600 nanometers — hey, that's not just microscopic but *nano*scopic! Yep, most of the light you deal with does its thing on a nano scale. We're not saying that photons per se are nano-size (remember, they don't really have a size), but their *vibrations* are nano-size. And it turns out that any nano-crystal router worth its salt — the kind of routers set to kick the speed of information transfer up a notch or two — is going to have to use the wavelengths of photons as a way to identify them.

You identify wavelengths all the time when you listen to any of the latest hip-hop tracks on the radio — or even Pink Floyd on that '70s station you'd never admit to listening to. Short-wavelength sounds make for the high-pitched vocals, and long-wavelength sounds make for a thumping bass line. Now, it may seem weird, but a nanoscopic crystal has its own "ear." It identifies different wavelengths of light by responding to *how* they travel. It turns out that different wavelengths of light travel at different angles when they're passing through a medium, separating into their signature colors (a process called *diffraction*) — as you can see for yourself if you hold up a prism and let the light shine through and fan out. What you get is that wonderful rainbow pattern that thrilled everybody back in high-school physics class. (You can also use a diamond, if you have one lying around — a soap bubble also works in a pinch.) The rainbow created this way also clearly identifies the various wavelengths of light: Each wavelength gives off a different amount of energy — which the human eye perceives as a color.

Optical communication is a pretty exclusive nightclub; usually we only allow 1500-nanometer wavelength light (part of the infrared spectrum) into the party because it's the telecommunications standard wavelength. So, if our crystal-based router is specifically designed to the 1500-nanometer wavelength, it can be integrated into the Internet. It's important to note that different materials and crystal designs can be specifically tailored for a specific wavelength and only this specific wavelength, excluding all others. If you don't have the 1500nm ID, you can't come in.

Controlling light: Photonic band gaps

Photons and electrons don't have a lot in common — photons are "light" (massless) and fast and electrons are heavy and slow and never the twain shall meet — but similar technology is needed to manipulate each of them. That's no fluke; when you replace a slow electrical device with a quicker optical one, you often use the old design to generate ideas for the new one. For example, old-school semiconductor manufacturers "dope" silicon crystals with just enough impurities to dramatically change the crystals' electrical properties. We can borrow that technique; if we add just enough of the right impurities to our nanocrystals, we get semitransparent structures: Some light won't pass through, but some will — and come out in a specific range of wavelengths. And as everybody knows who read the previous section of this chapter, different photons with different wavelengths have different levels of energy. By varying the geometry of the nanocrystal, you can change which energies get stopped by the opaque part of the crystal — and which ones pass through the transparent portion. You decide which photons to let out of prism (so to speak).

Want a great excuse to play with imaginary marbles? (No, we haven't lost ours.) Imagine a large bowl with the marbles spinning around inside. The fastest-moving marbles spin around near the top; the slowest marbles spin near the bottom, and most of the marbles spin somewhere in-between. Now put some putty in a circle around the middle of the bowl. Marbles that want to spin around at that level either get stuck in the putty or bounce to higher or lower parts of the bowl — and the marbles already in the high or low sections can't cross the putty. That putty functions as a *band gap:* It separates the marbles into groups that have different levels of energy, and keeps them separate (as conveniently illustrated in Figure 7-1).

Here's another way to think of this light-controlling stuff: We're interested in making materials that act as "light insulators" — materials that control how much (or what kind of) light is allowed to pass through the nanocrystal. These light insulators provide a *photonic band gap* that corresponds to a specific wavelength of light — photons that have that particular wavelength have to travel within the photonic band gap, restricted from the surrounding material.

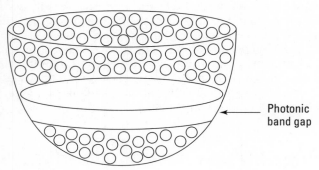

Figure 7-1: The energies of the marbles above the band gap are greater than the energies of those below the gap.

Photonic band gap

We can create band gaps in these "light insulators" (well, we might as well go ahead and call them *photonic crystals,* a specialized nanocrystal made to insulate light) one of two ways: either by exploiting geometric abnormalities in the crystal (missing or odd-shaped parts of the structure), or by using impurities in the crystal (the other materials we mix in). Additionally, changing the geometry of the impurities' molecules allows us to set up a couple of distinct energy levels inside the band gap — kind of like cutting a few tracks in the putty for the marbles to travel in.

Some photonic crystals are grown the same way computer chips are made. Those of you who manufacture your own chips in your garage (don't we all?) know that a silicon crystal is grown slowly, layer by layer, and impurities or geometries are etched in. You can create a honeycomb pattern in the crystal by etching opaque circles into each two-dimensional layer, spacing them at regular intervals. As you build up the layers with the circles always in the same places, the circles become rods.

Now, in some places, when we build our layers we don't include the circles, which in turn don't produce rods — just empty spaces. The chemical properties of these rods — when they are there — shape the photonic band gap: Light won't travel through a rod, but will travel in any space where a rod is removed. Remove enough rods, and the light travels along a nano-size path very happily.

To visualize our structure, think of these rods as jail cell bars scattered at regular intervals throughout a large room, as shown in Figure 7-2. You can only travel through the room where bars are removed. Remember, we are insulating light of a specific wavelength to be guided through our nanocrystal. The fact that we're using nano-sized defects allows us to control light (which operates in wavelengths of nanometer scale).

Bumper-bowling with photons

Remember digging canals in your backyard so your newspaper ships could get from Point A to Point B? (Remember what Dad did when he saw what you had done to his lawn?) The same thing is happening with the band gaps you etch into photonic crystals: The photons are traveling in the band gap where a rod is removed and only traveling there — they cannot cross into the rest of the crystal. This means the light will continue to travel along that path — with *zero* photons lost — no matter how that path twists and turns.

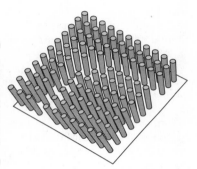

Figure 7-2 shows a top view of our photonic crystal. When the light approaches the turn in the crystal, it bends to follow the path. This is different from *diffraction* — the apparent bending and spreading of waves when they meet an obstruction. The light isn't traveling in a new material, and it's not changing speed; it's just changing direction — *at the speed of light* and *with no loss.*

Nanolasers and — I'm sorry, could you repeat that?

Along with routers, modern communication systems also need *repeaters* — in effect, in-line amplifiers that take fading light signals and resend them with more power. To do that with data shot along the line as well-ordered groups of photons, those repeaters need to be miniature coherent-light sources: nano-size lasers.

The photonic band gaps created inside photonic crystals not only provide an excellent way to keep the photons moving along certain paths, they can also provide areas that trap photons — *optical cavities.* As light enters such a cavity, the photonic band gap (etched into the crystal) keeps the light from leaving through the rest of the crystal. Such "trapped" light bounces back and forth in the cavity — gaining in energy, tightening into a coherent beam.

Certain materials — some semiconductors for example — can be stimulated to emit photons. The LED (*light-emitting diode*) is a common example of this kind of technology, but is not a laser itself. To get the laser effect, you'd most often want to choose a semiconductor-like material — called a *gain medium* — and place it in an optical cavity. Photons enter the cavity, bounce back and forth, and stimulate the medium to emit more photons. Those extra photons are of the same wavelength as the original ones, so what you get is an amplified version of the original light. (This "*light amplification by stimulated emission of radiation*" is how a *laser* gets its name.)

To make a nanolaser, a photonic crystal is used to create a cavity that's almost as small as the wavelength of the photons themselves. This cramped space forces the photons to travel in nearly parallel lines, until the intensity of the light reaches the theoretical limit — in effect, all the photons are traveling right on top of each other! The gain medium is essentially part of the crystal itself — but before it can emit photons, a small electric current must be introduced (that's true for most semiconductor lasers anyway).

That little zap of extra energy is all the photons need to make a break for it and blast out of the crystal as a laser beam. When supplied with a little electricity and a signal composed of photons, the laser amplifies that signal. Out comes a flow of happy photons, ready to spread information around the world.

Korean researchers have developed such a photonic crystal laser, using a semiconducting material (indium gallium arsenide phosphide). Their laser produces detectable amounts of light with as little as 250 millionths of an ampere of electricity. Their design uses a tiny post to conduct electricity and soak up excess heat without disturbing the main portion of the crystal at the top.

Self-assembling crystals

Part of the allure of nanotechnology is the ease with which many nano-structures can be made by a judicious use of chemical mixing — a pinch of this and a dash of that — rather than by using other more involved and lengthier procedures. Growing crystals, as mentioned in the "Controlling light: Photonic band gaps" section earlier in the chapter, is just such a more involved and lengthier procedure, which makes it a prime candidate for a bit of nano mixing-it-up.

One alternative involves a mixture of particles in a liquid referred to in chemistry as a *solution*. (Yep, the solution to our problem is a "solution.") In this mixture, submicron silica spheres float around in the liquid (called the *solvent*), which is most often ethanol. Ethanol is a type of alcohol, the same as found in liquor, which may explain why the little spheres tend to crash into a plate that is placed in the solution. As the temperature rises and the ethanol evaporates, the *meniscus* (a fancy word for "the surface of the liquid") travels down the plate, making it easier for any submicron silica spheres floating nearby to stick to the plate. As time goes by, more and more spheres deposit themselves on the plate, forming an orderly pattern. When all the ethanol is evaporated, the first layer is finished and subsequent layers can be added by repeating the process. Figure 7-3 illustrates this process — which is (as enquiring minds would like to know) called *colloidal self-assembly* — *colloids* being the solution of silica spheres (between 1 nm and 1 micron in size) suspended in a solvent.

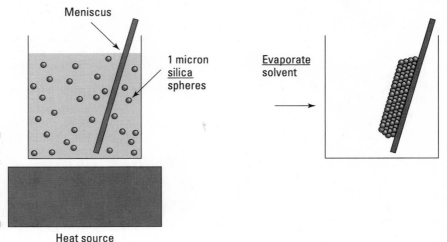

Figure 7-3:
Diagram of the self-assembling process.

By varying the size of the spheres being placed in the solutions, you can introduce useful chemical defects at specified layers.

After we have all the spheres lined up (see the left side of Figure 7-4), we coat the spheres in a polystyrene plastic — we even get it into the dead spaces between spheres. Once that hardens, we use a chemical etching process to remove the silica spheres, which produces an inverted representation of the original structure (as shown on the right in Figure 7-4) — the *superlattice* we're looking for.

Making these crystals is very quick and (relatively) easy. Unfortunately, they also don't perform nearly as well as their slow-grown silicon-crystal counterparts — we can't control the defects as easily or precisely. Researchers in both groups — the self-assembly types and the homegrown types — are trying to improve their techniques so that someday all of us nanotechnology aficionados can have the best of both worlds.

Figure 7-4:
The crystal
(left) after it
assembles
itself, as
seen
through a
scanning
electron
microscope
(SEM). The
inverted
nanocrystal
(right)
containing
the photonic
band gap.

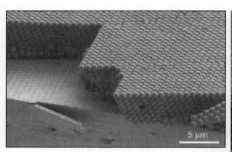
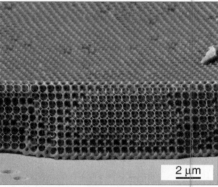

Courtesy of Nature [www.nature.com] and David Norris from the University of Minnesota

Optical switching: Nano-defects to the rescue!

Being able to form different paths for photons without being able to switch between those paths is sort of like having guest towels without guests: Looks nice, but who gets to use them? You have to be able to "flip the switch" (think electric train sets here) in order for a nano-optical routing device to work. In order to come up with a way of doing that, we need to take another look at those photonic band gaps as well as the ways in which photons like to move around, and maybe we can find a way to get switching to occur — otherwise our nano-routers are out of luck!

Picture, if you will, a bowl full of marbles spinning around — yep, the example we plunked down at the beginning of the "Controlling light: Photonic band gaps" section earlier in the chapter. The putty that separates the top and bottom halves of the imaginary bowl doesn't necessarily have to form a full circle. There *can* be small holes that allow marbles to roll through the putty layer, from one half to the other. A really fast marble, rolling discontentedly around the bottom half, would roll along the putty until it came upon a hole, and then it's outta there — up into the top half it goes. Now, if there were *two* lines of putty, spaced a bit apart from each other, and holes in each of those lines, a marble could conceivably roll around until it came to a hole, pass into the space created by the two lines of putty, and then roll around until it came upon another hole, ending up once again (voilà!) in the other half of the bowl. All of this is brilliantly illustrated in Figure 7-5.

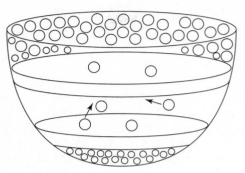

Figure 7-5:
Analogy of a photonic band gap, with the multiple allowed energies that can travel inside.

"Okay," you say, "thanks for that intriguing discussion of bowls and marbles, but we want to learn about nanotechnology now, please." We don't mind cutting to the chase: The photonic crystal can be structured with certain defects inside — on purpose — in order to channel photons within the crystal. Fortunately, such flaws are pretty easy to create: We just remove some rods (as we discussed earlier) or add in some specific impurities. With the defects in place, we no longer have just a simple photonic band gap. Instead, we have a band gap that allows a few photonic trespassers across the border. (This is the same idea as the two rings of putty with some space in the middle, which we describe in our bowl-and-marbles example.) Photons still can't cross *directly* from one side of the gap to the other, but they can go to the intermittent level and *then* to the other side. "Oh," you say, "now I see where you were going with all that bowl-and-marble stuff." Thanks, gentle reader, for your patience.

Intermittent levels allow photons to cross the photonic band gap — they're the little wrinkles or folds that let photons "cross over." Most of the time, these intermittent levels are hard-wired into the crystal, but sometimes they can be manipulated by outside energy sources, such as additional photons. If you could regulate those photons, you could make an optical switch that you could turn on and off by shining a light at it. (Now, I've heard of a switch turning on a light, but a *light turning a switch* off and on? Why not?)

Making the switch: Photons on a nano-highway

Researchers at Cornell University have already developed some of the nanotechnology needed to create an all-optical switch — they solve the problem of using light to switch light. Using just picojoules of energy, they can switch paths in a few hundred picoseconds of time. (*Pico* is the prefix meaning "one-trillionth," so this is both incredibly fast and also incredibly sparing in terms of energy usage — one AA battery has about 1000 joules of energy.) Even more impressive, Cornell's nanocrystals use the most readily usable material — silicon. Even better results have been obtained elsewhere using more exotic substances, but silicon is the best choice if you want to mass-produce nanorouters.

Ready for a segue to another homespun metaphor? If you've ever driven around in Europe (or in the northeastern United States), you may have come across one of transportation's most useful — and sometimes dangerous — tools: the rotary. (The Brits call them "roundabouts.") A rotary is little more than a circle of roadway connected to several incoming roads. Drivers can enter the rotary from a feeder road, travel around in the circle, and then exit onto another road, all without having to stop.

The photonic equivalent (minus the fender-benders) is called a "ring-router" and is pictured in Figure 7-6. Photons can travel along an attached path until they come into the ring router and start zipping around at breakneck speed. Unlike some drivers from New England I'd like to mention, the photons never ram into each other or honk like maniacs.

Figure 7-6:
A picture of a ring router, taken by a scanning electron microscope (SEM).

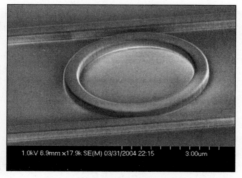

1.0kV 6.9mm x17.9k SE(M) 03/31/2004 22:15   3.00um

Courtesy of Michal Lipson from Cornell University

The ring router has intermittent levels in its photonic band gap (in terms of our silly bowl of marbles, *several* lines of putty) — and we can manipulate the levels by introducing photons of the proper wavelength. When the right beam of photons comes along, the router changes its task: Instead of allowing light to zip around in it and go through it, it absorbs the new wavelength or passes it off in another direction. Figure 7-7 shows how such a switch might work. Notice that there are now *two* beams we have to keep track of — the signal beam (the one that either gets blocked or passed through the router) and a switch beam (the one that turns the ring router on and off).

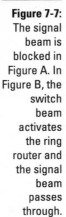

Figure 7-7:
The signal beam is blocked in Figure A. In Figure B, the switch beam activates the ring router and the signal beam passes through.

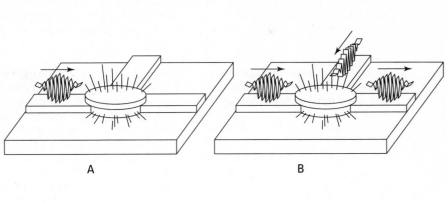

A   B

Hey, when we get to that point, we have our nanorouters. That was quick! We're going down to the store and snap up one of those all-optical switches right now. Wait a second, where are all the optical switches? Oh, that's right . . . nanocrystals as described in this chapter can only rout nano-size portions of light. That's fine for small sections of the information superhighway — but if we want the equivalent of a 20-lane cloverleaf intersection, we're going to need something larger. . . .

Magic with Mirrors

For many children, as well as for many graduate research assistants who seem to think they still are children, magicians can entertain and induce wonder seemingly without end. Forks and spoons can disappear underneath a deftly maneuvered napkin, sharp swords refuse to cut soft skin, and money springs forth freely from behind our ears. After the finale of the show, however, it is all revealed to be smoke and mirrors — a series of illusions and tricks.

It may not be magic, but nanotechnology offers to reproduce many of the same feats — for real. Nano-designed catalysts and reactants (for example) can help break down previously non-biodegradable materials and make them disappear from the trash dump. Composites of polymers and nanofibers could make super-strong fabrics to stop bullets and shrapnel. And we all hope that these inventions produce an endless supply of cash.

Yet, when dealing with light, nanotechnology still resorts to smoke and mirrors — well, mirrors anyway. Mirrors are a handy, easy way to direct light around — and by doing so (in this case), to help control the information superhighway. Reflecting large portions of a light beam is pretty easy: Just hit the mirror at the correct angle and watch the light fly off at the same angle.

When you change the direction that the mirror faces, you change the direction that the information flows — it's routing in bulk! While it doesn't get the details sorted out, it does move a *lot* more data around than the photonic crystals could by themselves. Combining the two techniques could provide a more well-rounded and versatile router.

But how do we steer those mirrors? Enlist some little nano-size men to the cause? Interesting, but a tad impractical (*Fantastic Voyage* was fiction, after all). We'll probably have to settle for one of the many other ways to move mirrors — to *beam-steer,* as the true aficionados refer to it. Figure 7-8 shows a fairly tiny mechanical solution: 256 mirrors on a few square centimeters of silicon — about the size of a postage stamp.

Using electrical pulses to move the mirrors provides even finer control — and allows faster beam-steering. Put enough of these fast-steering mirrors together and not only do you get one heck of a router, but also whole new ways of recording images. Such nano-imaging can even be *active* — the optical elements (such as the mirrors) move themselves to optimize what they are seeing. Nanotechnology is providing so many new ways to use mirrors — it's starting to look like magic.

Harnessing the force . . . of nanotechnology

Not only can these mirrors record images, they can also project. A famous example appeared in May 1999: the first episode of the *Star Wars* movie series. It was the first movie to use a Texas Instruments technology called *digital light processing (DLP)* to project the images onto the screen. DLP is based on the optical semiconductor called the *digital micromirror device (DMD).* We did say "micromirror" — although small, micro is a thousand times bigger than nano. Each DMD chip contains an array of 2.1 million mirrors — each 16 microns square, capable of switching 5,000 times per second and displaying 16.1 million colors. It takes only one chip to produce images for a DLP television — and in DLP cinema, they use three: one for each color (and such a system can produce 35 trillion colors).

Although this falls into the bailiwick of the micro-electromechanical systems (MEMS) discussed in Chapter 8, DLP is one of those technologies that can benefit hugely from nanotechnology (as in: tiny components, big impact). A possible application is one we describe in Chapter 6: holographic storage. InPhase Technologies, for example, uses DLP in their holographic storage system — and that's only the beginning. Nanotechnology has already shown that it can bridge gaps between industries — from routing to entertainment to high-density storage.

Figure 7-8:
At top: a micro-size mirror allowed to tilt in any direction. At bottom: A conceptual diagram of the optical routing that the mirrors provide.

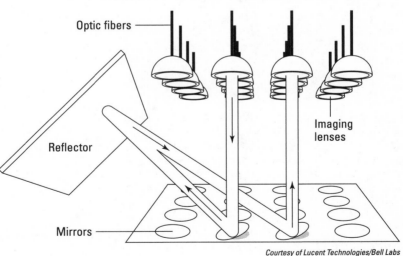

Optic fibers

Reflector

Imaging lenses

Mirrors

Courtesy of Lucent Technologies/Bell Labs

Light-steering: Nanotechnology at the wheel

Aiming a mirror at the nano scale requires (surprise! surprise!) precise control of tiny electromechanical devices. Pushing or pulling on a mirror adjusts the angle at which light bounces off; even small changes in angle can lead to big differences in the final destination of a beam of light. To tightly regulate the angle of a mirror on the nano scale, you have to spread out supertiny electromechanical devices around its base. Each one pushes up and down on its portion of the mirror like a . . . little nano-size man. (Hmm . . . maybe we ruled out that option a bit too soon.)

Perhaps the most common form of electromechanical device is the *piezoelectric transducer* or PZT — typically like those disks you see sketched out in Figure 7-9. A PZT is built from a material — almost always a ceramic — that expands and contracts according to the amount of electric current that travels through it. (Unfortunately, PZTs often twist or pitch as they expand, limiting their precision. Even so, when you control it with a very precise current supply, the PZT can have submicron resolution.) Stacking multiple PZTs creates a greater dynamic range — labeled ΔL in Figure 7-9. Whether you use a stack or single PZT, however, a simple linear setup is always going to be limited by two things: the quality of your supply of electric current and the amount of twisting the PZT undergoes.

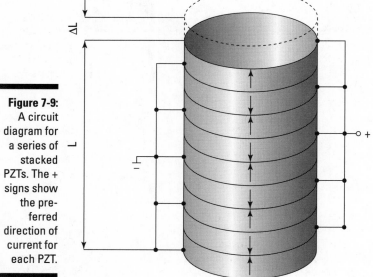

Figure 7-9:
A circuit diagram for a series of stacked PZTs. The + signs show the preferred direction of current for each PZT.

For the next level of precision, you have to go to *flexures,* which have more aliases than your average super-spy. They are also called *flexors, flexing actuators,* and *deforming nano-positioners.* Essentially, the flexure concept means that flexible joints have been added to either one PZT — or, more often, a team of PZTs — to counteract the twist that occurs as PZTs expand. (See Figure 7-10 for a diagram.) The size of such a setup is really only limited by the current supply of components; theoretically (ahem) it could be infinitely small. Sources of electricity, however, are not perfect, and there will always be noise and drift in the amount of current you can provide. Still, it's already possible to achieve precision of less than one nanometer for flexures.

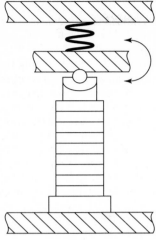

Figure 7-10:
A basic PZT and flexure arrangement, where a ball joint, platform, and spring compensate for the undesired twisting of the PZT.

Although it's long been the premier technology of its kind, a complex PZT system is also very expensive and difficult to produce — and a nano-scale PZT positioning device can cost a million dollars without working up a sweat. Many companies are advancing production designs and materials to bring that cost down; they're also designing newer and more precise versions for ultrasharp control. Before you go sticking a mirror on the end of one of these systems, however, you really should consider another option.

Researchers at Boston Micromachines Corporation (BMC), in association with Boston University, have refined a technique to position mirrors on the nano scale. Instead of using materials that contract or expand with electricity, they are using electric attraction to push and pull on the mirror itself. Although this is a more limited range of motion, it can be very precise.

For an everyday-scale demonstration of the strength of this type of attraction, just rub your head with a balloon on a dry day (if anybody looks at you funny, just say it's for science). Your hair will slowly cling to the balloon, and try to stay there as you pull the balloon away. Place the balloon against the wall and you will see it stick there until it slowly falls down (on a really dry day, it may stay on the wall without falling). This attraction is familiar to us as "static cling" and "static electricity," but if you want to sound scientific, call it *electrostatic force.*

The inventors at BMC use thin layers of silicon and oxide to stick a mirror to a tiny platform. Then they dissolve the oxide; the result is a mirror with a honeycomb of silicon attached to its base. As voltage is placed across that base, electrostatic forces attract the mirror across the gaps of the honeycomb (which are about 5 microns across). Controlling that voltage is the work of some complicated software, but in the end, you have a mirror that can be pulled with nanoscopic precision — not just in one place, but in thousands of places.

Figure 7-11 shows a finished BMC mirror — sometimes called a *deformable* mirror for the way it can change its shape. (It's the smallest reflective square in the center of the circuit board.) Resolution in positioning is a few nanometers, and improvements are underway. For now, it's a tradeoff: Devices that use electrostatic forces are as precise as those driven by PZTs but don't provide as much displacement; the costs, however, are dramatically cheaper — as little as one-eighth the cost of comparable PZT technology.

No matter which type of mirror-positioning device you'd like to use, the final step in including it with the nano-router is fairly simple. (Remember: The whole idea here is to use the mirror to do direct information in the form of light, thus speeding up the Internet for us.) You can either attach a mirror to the tip of the positioning device, or you can shine light at the tip of the device so it reflects the appropriate wavelength (although opportunities to use the latter technique are still limited). In the end, the nano-router's use of mirrors will probably be limited to bulk transmissions of data; photonic crystals are faster for smaller chunks of information. But, hey, just remember that every highway needs its 18-wheelers as well as its zippy sports cars.

Figure 7-11: But can you use it to mail a letter? The BMC's precision mirror device, as packaged for sale, compared to a stamp for scale.

Courtesy of Paul Bierden of Boston Micromachines Corporation (BMC)

Mirror, mirror on the wall! Nano's the sharpest image of all

Now that you can send information around the world at the speed of light, maybe we should start gathering some of that info. Nanotechnology is starting to take a close look at processing visual images to greatly improve the information they can provide us. *Active imaging,* the process of adapting to flaws in images while you look at them, has been around for while. Having your pupil dilate when a light gets too bright is a form of active imaging. So is focusing the lens of your eye on something far away. But let's face it; dilating pupils just doesn't cut it in the Excitement Department. To get the pulse racing, you really have to go nano.

In Figure 7-12, you can see three versions of the same image: Titan, Saturn's largest moon. Now, due to the (relatively) poor quality of their lenses and the disruption of the earth's atmosphere, normal telescopes show us little more than a nice gray blob. This may have been good enough for Galileo, but not for the folks who launched the Hubble Space Telescope. Without the distortions of the atmosphere — and, admittedly, with better lenses — the Hubble telescope shows some marked improvements. If you're thinking space is the place for seeing distant objects up close, think again. The Keck telescope — an earthbound observer — receives the best image. Why? Chalk it up to adaptive optics (AO) — a marriage between nanotechnology and computer science.

Figure 7-12: Saturn's largest moon seen through (a) a conventional telescope, (b) the Hubble Space Telescope, and (c) the Keck telescope with adaptive optics.

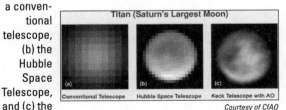

Courtesy of CfAO

Those little deformable mirrors we were going to create to route information at the speed of light have some other uses, one of which is adaptive optics. As light from the sun bounces off of Titan and heads toward Earth, it encounters all sorts of interference, be it dust, the earth's atmosphere, or little green men. Eventually the light reaches a telescope, with its own aberrations and distortions, and you or I can see what's happening on that side of the solar system — if we can see past the distortions. Now, suppose you sent a copy of the image to a computer, bounced another copy of the image off our little deformable mirror, and recombined them in a way that cancelled the distortion. Pretty slick. That's the basis for an adaptive optical system.

Computer scientists — "nerds" to those who know and love them — have come up with some pretty amazing algorithms to check how an image is being distorted. Too many bright pixels in one area may be a virtual light source; too many wavy lines may be some sort of astigmatic aberration — and a big fly-shaped shadow may mean we need to stop leaving pizza out all night. Whatever the case may be, good computer software is available to analyze an image and help correct it. (You may have seen this sort of program in action on cop dramas where they need to "enhance" a video image to catch some bad guy.)

Now, when you get the computer to control the nano-routing/deformable mirror, it can bend and flex it in a thousand different spots. By bending in the right shape, it can redirect the distorted light back to where it should go — it actively compensates for the distortions in the image as they happen. Furthermore, as the computer gets immediate feedback as to whether or not the compensation is working, the overall improvement in image quality increases. Figure 7-13 helps lay out this adaptive-optics process visually. Forget cop dramas — this is more like *Star Trek*.

Adaptive optics becomes very helpful when trying to resolve nanometer-sized objects. In many cases, the size of the object may be *smaller* than the wavelength of light you are using to shine on it. This is like trying to pick up a penny with boxing gloves — no one's getting rich that way. Any additional resolution that the adaptive optics can provide is going to help.

Try looking at it through nanotechnology's eyes

When you come right down to it, no matter what sort of imaging equipment you have, you eventually have to use the good ol' human eye to see what you've got. For many of us science geeks, that means glasses . . . contact lenses if you're a chic nerd on the go. So improving human eyesight seems only fair after we've spent all this time making telescopes and microscopes have better vision.

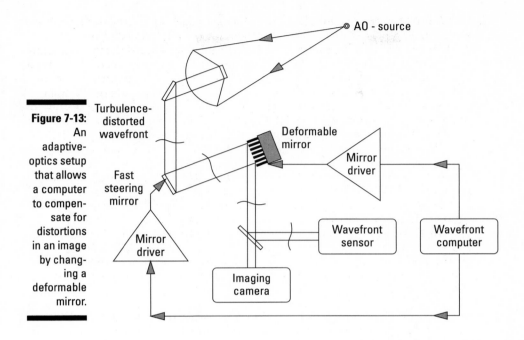

Figure 7-13: An adaptive-optics setup that allows a computer to compensate for distortions in an image by changing a deformable mirror.

Most researchers in ophthalmology will tell you that looking at the minute details of the human retina (the fleshy bit at the back of the eye) is limited by the distortions caused by the eye's lens. With the help of the deformable mirror and adaptive optics, however, we can compensate for such distortions and produce crystal clear pictures of the retina.

The University of Rochester's Center for Visual Science has put to good use some of the deformable mirrors from BMC we talk about in the "Light-steering: Nanotechnology at the wheel" section, earlier in this chapter. They call their baby an Adaptive Optics Scanning Laser Ophthalmoscope (AOSLO) and use it to study the retina and help find cures for vision disorders. Similar work has also been performed at the University of Houston, and other universities around the world.

Additionally, adaptive optics may revolutionize all sorts of sporting and military optics. If the computer components were adapted, AO technology could work on binoculars and similar equipment. The aforementioned universities have already shown that an observer's recognition of shapes and orientations improve when looking at AO-enhanced images. Everybody from Army snipers to birdwatchers will be able to get a better idea of what they're looking at. Who knows, maybe my dad will finally be able to spot that elusive blue-ridged speckled southern house wren . . . or Bigfoot.

Chapter 8

Nano-fying Electronics

. . . .

In This Chapter

▶ Generating light with nanotechnology

▶ Sensing things with nanotubes

▶ Building MEMS — those teeny-tiny electromechanical machines

▶ Making computer brains one molecule at a time

▶ Connecting devices with nanotubes and nanowires

. . . .

Consider the world of electronics — you know, that world that provides us with light at night, enhanced abilities to sense things around us, and a way to run complex machines. Electronics also makes possible the computer that you probably sit in front of almost every day. What if all these things could be driven by nanotechnology? That might improve the efficiency of — well, just about everything — and lower the cost of our daily lives quite a bit.

This chapter is about providing the world of tomorrow with faster, more powerful, and more efficient electrical devices, all driven by nanotechnology.

Lighting Up Tomorrow

Edison made a giant leap when he lit up the wire filament in that first incandescent light bulb. He was the inventor-hero of his day, but Edison was working in the Dark Ages compared with the things that nanotechnologists are exploring today. The world may just become a brighter place when we start seeing quantum dots before our eyes, and fire up light-generating nanotubes too small to see.

Making quantum leaps with quantum dots

Quantum dots are nano-size crystals that emit light; the wavelength they emit depends on the size of the crystal. Quantum dots are composed of various materials, such as lead sulfide, zinc sulfide, cadmium selenide, and indium phosphide. Quantum dots are useful because, depending on their size and composition, they emit a particular wavelength, or color, of light after an outside source, such as an ultraviolet light, excites the electrons in them. Quantum dots produce light in a way similar to atoms, as we discuss in the next section. The ability to tailor the color of light emitted by a group of quantum dots is very useful in medical diagnostics, as we discuss in Chapter 10.

The rules that describe electron orbitals (also called energy levels) — and dictate that electrons are only allowed to be in certain energy levels within an atom — are called *quantum mechanics*. Because electrons in these nano-size crystals behave in a similar way, they are called quantum dots. (For more about the nano-effects of quantum mechanics, see Chapter 6.)

Getting quantum dots energized

Quantum dots are useful because when you add energy to their electrons, the electrons act like they're in one big atom — and (as any physicist could tell you) when you add energy to the electrons in any atom, what you get is light. This occurs when an electron moves to a higher energy level and then falls back again to its normal energy level. The same is true for quantum dots — zap them and they glow. One way to add energy to quantum dots is to shine an ultraviolet light on them.

It turns out that the smaller the quantum dot, the larger the gap between energy levels — which means more energy is packed into the photon that's emitted when an electron falls from a higher energy level to its normal energy level. A small quantum dot emits higher-energy photons — with a shorter wavelength — than a large Q-dot can.

Think of this light in terms of color: A quantum dot of a particular size — a relatively large size, to be exact — emits red light, which is the longest wavelength of visible light; smaller quantum dots produce different colors. If you keep going down smaller and smaller, you'll eventually get to a tiny quantum dot that emits blue light — the shortest wavelength of visible light. If you come up with *really large* quantum dots (okay, we're being a *little* silly here), you might get them to emit infrared light; incredibly teensy quantum dots might emit ultraviolet light, outside the visible spectrum.

So where do you get quantum dots? (No, you can't find them at Wal-Mart — at least not yet.) It turns out that it's possible to grow a large number of quantum dots in a chemical reaction — but the methods used range from simple wet-chemical setups (in which you precipitate zinc sulfide crystals) to complicated methods such as chemical-vapor deposition (which is also used to grow carbon nanotubes, as we discussed in Chapter 4). You can control the size of a particular batch of quantum dots — ensuring that they all emit the same wavelength of light — by controlling the length of time you allow the reaction to run. But what do you do with them once you've got 'em?

Putting quantum dots to work

Quantum dots — aside from being neat things in and of themselves — could be put to some very interesting uses. The first application of quantum dots will be for biological labels used in medical imaging. Researchers tag proteins and nucleic acids with quantum dots; when they shine ultraviolet light on a sample, the quantum dots glow at a specific wavelength and indicate the locations of attached proteins. Quantum dots have advantages over materials currently used for this application — for example, they glow longer. (For more about the medical applications of quantum dots, flip over to Chapter 10.)

Researchers are also hoping that quantum dots could eventually provide energy-efficient lighting for general use — in your home, office, or neighborhood street lamp. In these applications, a light-emitting diode (LED) or other source of UV light would shine on quantum dots, which would then light up. By mixing different sizes (and the associated colors) of quantum dots together, you could generate white light. Generating light from quantum dots would work like generating fluorescent light — but without that bulky fluorescent tube. This method would also avoid the wasted heat that you get with your typical incandescent light bulb.

Passing an electrical current through an LED also generates light. A company called QD Vision is attempting to use techniques developed at MIT to design a quantum-dot LED: a layer of quantum dots sandwiched between conductive organic layers. Passing a current through the dots generates light (as shown in Figure 8-1). Flat-panel TV displays using quantum-dot LEDs may provide more vibrant colors than current flat-panel displays based on Liquid Crystal Display (LCD) technology.

Figure 8-1: Structure of a quantum-dot LED.

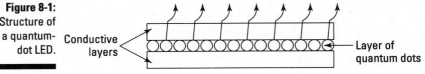

Light emitted from quantum dots when electric current is passed through them

Conductive layers

Layer of quantum dots

Getting light from nanotubes

Researchers at IBM, not to be left behind, have found that when you flow electrons through semiconducting carbon nanotubes, the nanotubes emit light at specific wavelengths — a nano-scale version of *electroluminescence* (converting electrical energy into light). Figure 8-2 shows an example of an IBM-style field-effect transistor (or FET, discussed in more detail in Chapter 6) in which electrons flow through a nanotube to produce light.

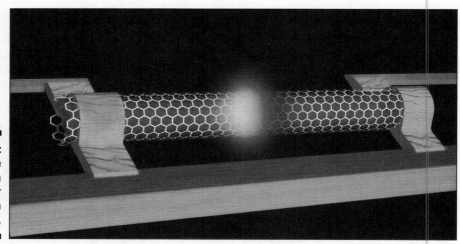

Figure 8-2:
A nanotube used as a channel for electrons in an FET.

Light is light, no matter where it comes from — and if a nanotube generates light at a specific wavelength, it's similar to the light generated at the same wavelength when electrons move between these two states (each with its own energy level) in an atom or molecule:

✔ **Conductance:** An electron in a *conductance state* is free to move around from atom to atom in the nanotube. Electrons in a conductance state have a higher energy level than those in a valence state.

✔ **Valence:** An electron in a *valence state* occupies an orbital in the outer shell of an atom in a nanotube, at a lower energy level than those in a conductance state.

Now, remember, electrons tend to be lazy; an electron with more energy wants to get rid of it somehow. That's why an electron kicks out a photon when it settles down from a conductance state into a valence state. The difference in energy between the two states is emitted as light.

The energy level of an electron in a carbon nanotube is determined by the diameter of the nanotube; as with quantum dots, a smaller diameter means a higher energy level. Therefore, the wavelength of the light generated also depends upon the diameter of the nanotube. Think of a guitar: Each string has a different diameter, and emits only one certain note. In a similar way, only a certain wavelength of light is emitted by each size of carbon nanotube. In effect, you can "tune" a nanotube to emit light at the energy level you choose. Researchers believe that optoelectronic devices — say, the amplifiers used in fiber-optic circuits by the communications industry — could be built making use of this effect. (At the moment, nobody's figured out how to use this effect to play "Stairway to Heaven.")

Sensing Your Environment

In essence, the world is a great big place made up of very small particles. It makes sense that going to the nano level will help us discern the finer-grained details of the world around us. From sensing chemicals to sensing biological molecules, nanotechnology is hard at work.

Detecting chemicals

Scientists, in their geeky wisdom, have developed sensors that use carbon nanotubes as sensor elements to detect chemical vapors — even if said vapors are scarcely present, with concentrations measured in *parts-per-billion* (ppb, also known as *really, really small amounts*) — within mere seconds. (So just *how* do nanotubes actually do such remarkable things? You'll have to read the next sections to find *that* one out.)

It's a chemical reaction

Remember those nasty chemical smells you had to endure in your high-school chemistry class? Besides making you gag, chemical vapors can also change the conductance of a carbon nanotube when a molecule bonds to the surface. This bonding can either increase or decrease the amount of electrons available for conductance in the nanotube. That's how some nano-based sensors work — they detect the change in conductance when a semiconducting carbon nanotube comes into contact with certain chemicals.

For example, nitrogen dioxide (NO_2) reduces the number of electrons available in a carbon nanotube. In this molecule, the nitrogen atom is one electron short of becoming a noble gas — which, to be honest, is high on the list of what it would just give its eyeteeth to do. (Keep in mind that a *noble* gas isn't part of some inbred, Eurotrash clique of titled aristocracy; rather, it's a gas with the most stable number of electrons of any element. See Chapter 3 for more about noble gases.) When NO_2 bonds to the surface of a carbon nanotube, it filches an electron for its own use — which decreases the conductance of the nanotube.

On the other hand, ammonia (NH_3) combines with water vapor in the air to form a molecule that can *contribute* an electron to the carbon nanotube, *increasing* its conductance.

Chemical vapors that have been shown to change the conductance of a nanotube include nitrogen dioxide, ammonium, carbon monoxide, carbon dioxide, methane, hydrogen, and oxygen. Molecules that include these groups have also been shown to be detectable by nanotubes. For example, nitrotoluene — a chemical used in many explosives and containing nitrogen dioxide — has an effect on a carbon nanotube similar to that of nitrogen dioxide — so it may be detectable by the same means.

Such nanosensors work because atoms or molecules bond to the surface of the nanotube — and thereby change its electrical characteristics in telltale ways. As you might expect, the sensitivity of the sensor depends on how much surface area is exposed to the gas — and as miniscule things go, the surface of a nanotube is pretty big. Just as a grain of sand has more surface area relative to its volume than a huge rock has, nano-scale materials have a bigger surface area (relative to volume) than other materials used to sense chemicals. There's an even better reason, however, to use nano-scale materials for sensing chemicals: A very small amount of a chemical can change their conductance. So (on the molecular scale, at least) the relatively large surface area of the big honkin' carbon nanotube makes it sensitive to smaller traces of the chemical you're sampling. In the next section, we discuss how you sense these changes in conductance of the nanotubes.

Detecting chemicals in the real world

One interesting application of these nanosensors is the detection of trace levels of vapors from explosives. Because most explosives are made of molecules containing nitrogen dioxide (NO_2), nanosensors may be able to detect trace amounts of those vapors. This potentially greater sensitivity might eventually help airport security personnel detect (for example) bombs in luggage.

But the road to efficient nanosensors isn't without potholes. Before we can put these sensors into use, we have to solve one problem: It's not always easy to determine what molecule you've detected. That's because (in principle) any molecule that *can* bond to the carbon nanotube and contribute or remove an electron — even if it isn't the one you're looking for — could cause the sensor to respond.

For that reason, researchers are investigating how to "tell" the sensor what to look for by coating the nanotube with various materials. Any sensor based on a coated nanotube is responsive only to certain chemicals — for example, coating the nanotubes with a polymer called polyethyleneimine (don't even try to pronounce that one) allows the sensor to detect NO_2 but minimizes its sensitivity to NH_3. Coating nanotubes with a different polymer called nafion stops NO_2 from getting through — but allows the sensor to detect NH_3.

But how does the sensor measure the change in the nanotube's conductance once it detects a chemical? Well, a quick look at a simple device that uses carbon nanotubes as a sensing element (a *chemiresistor*) shows how: In a chemiresistor, carbon nanotubes are laid between metal electrodes; you apply a constant voltage between those two electrodes. If the conductance of the carbon nanotubes remains constant, the resulting electrical current also remains constant. If the conductance of the carbon nanotube — and therefore the current — changes, then a chemical is present that caused the change.

Chemiresistor variations

A group at UC Davis is experimenting with chemiresistors that use *nanowires* — very tiny wires made from metals (such as silver) or semiconductors (such as silicon) instead of the carbon you'd see in nanotubes — but produce similar results. Meanwhile, the Naval Research Laboratory is working on a chemiresistor-based chemical-vapor sensor that uses gold nanoparticles. The idea here is that molecules that intrude between the gold nanoparticles change the resistance that the sensor can detect.

Another method uses a field-effect transistor (FET), as described in Chapter 6. Using this method, only the top side of the carbon nanotube is open, and that's the only place chemicals have contact with the nanotube. (If you are familiar enough with transistors to ask where the gate of the transistor is, it's on the back of the device.) The conductance of the carbon nanotube changes as a result of contact with the chemical vapor — which in turn changes the gain of the transistor — and the chemical is caught in the act of being there.

The surface area — and presumably the sensitivity — of this type of detector is increased by constructing each electrode in the shape of multiple "fingers." You then interleave the fingers from the two electrodes and lay carbon nanotubes between the interleaved fingers, as shown in Figure 8-3. A sensor of this type can detect NO_2 levels of less than 50 parts per billion.

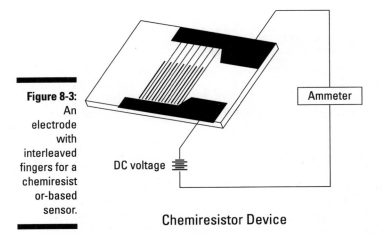

Figure 8-3: An electrode with interleaved fingers for a chemiresistor-based sensor.

Biosensors

If sensors can detect chemical vapors in the air, it's not much of a stretch to imagine sensors — designed along the same nanotechnological principles — that would be able to detect biological molecules or viruses in the air. In fact, we're on such a roll here that we're going to consider such variations small fry and move on to something really cool. How about creating biosensors that can be used in the bloodstream? Because you could use this technology to sense various medical conditions in people, this is an area of particular interest to medical science.

Paradoxically, it's a huge challenge to determine what causes a tiny change in the conductance of a sensor element introduced into the bloodstream. This is because many of the molecules, viruses, and proteins zipping through it are electrically charged — their molecules have either more or fewer electrons than protons. This charge can be transferred to the sensor element, creating *noise* — electrical interference — that makes it hard for the sensor to pick out the signal it's looking for. To solve this problem, we'll have to develop coatings or other methods that limit what can reach the sensor element. (Biosensors are discussed in more depth in Chapter 10.)

In his book *Prey*, sci-fi author Michael Crichton depicted a future with nanosensors running amok all over the place. But don't worry: Creating swarms of nanosensors to ruin your next picnic is still a ways in the future. Nanosensors are still a new development; they aren't cheap, but at some point their cost is going to come down. When it comes down far enough, we can indeed deploy these tiny machines in swarms. Then they can act in force, backing each other up in what scientists like to call a *higher level of redundancy* — which can make them much more effective.

Mechanizing the Micro World

A couple of centuries after the Industrial Revolution, our world is full of machines — ranging from simple to complex, from motors in cars to decision-making robots. Where does nanotechnology fit into mechanics? There's one little acronym that holds great promise in this area: MEMS.

Micro-electromechanical machines (MEMS)

A *micro-electromechanical system,* or MEMS, is, logically enough, a mechanical system or machine that exists at the micro level. To quantify sizes down there, the best unit of measurement is the micrometer ("*micro*-meter," not "mi-*crom*-eter," which is an everyday-size precision tool). This unit, often referred to as the *micron,* equals one millionth of a meter (about one hundredth the width of a human hair). Imagine machines built to the scale of microns. As with many more familiar machines, a MEMS might have gears, motors, levers, bearings, and so on that are capable of moving things — say, adjust a very tiny mirror to shine a pinprick of light on a flat-panel TV display.

If you've been paying attention to the details of the nano-size world, you may be asking yourself, why are they talking about *micro*-electromechanical systems, when they should be talking about *nano*-electromechanical systems? Well, for openers, right now nobody can *make* nano-scale electromechanical systems. (Details, details.) Remember, a nanometer is one billionth of a meter in size — way smaller than a micrometer. But it's quite possible that by using micro systems we can build smaller and smaller elements to help us reach the nano level some day.

How MEMS work

So how are people building MEMS today? Well, traditional manufacturing techniques that you might use to build a regular motor just won't handle micro-scale precision work. Instead, folks are rolling out the same techniques used to make computer chips. Using that approach, eventually we can make these devices in large quantities (potentially in the millions), at low cost, and with uniform performance. At the end of the manufacturing process, instead of having a chip that processes information, you have a mechanical chip — a nano-scale machine that performs an action.

It's entirely possible to have a MEMS chip that contains an information-processing component as well as a mechanical component. For example, you might create circuitry to analyze signals from a sensor on the same chip that controls a mechanical device.

How people are using MEMS

Even this early in the nano story, folks have applied MEMS techniques to a variety of sensors. One example is an accelerometer (a sensor that detects sudden changes in speed) — handy in applications such as release mechanisms for car airbags. The MEMS version of the accelerometer uses a silicon shaft that holds a weight. One side of a capacitor is built into the shaft. When the acceleration changes, the shaft bends, and this changes its capacitance. The electrical circuit in the chip measures that change in capacitance — and determines the amount of acceleration. This type of accelerometer has the necessary precision to trigger automotive air bags in a collision, and is already in use — at a cost that's about 80 to 90 percent lower than for conventional accelerometers.

Another MEMS application is already in extensive use in medical settings: MEMS-based blood-pressure sensors. As with standard pressure sensors, these measure the movement of a diaphragm built into the device — but they're also disposable. Such sensors can be used at a small fraction of the cost of more traditional pressure sensors — and the use-once-and-throw-away approach is often safer if the patient has a highly infectious disease.

But MEMS don't have to just sit around sensing things; they can also manipulate or move things around. One of the few commercial applications currently marketed concerns MEMS chips that contain an array of microscopic mirrors. These mirrors act as optical switches that produce the picture in many large-screen TVs. Each mirror is assigned a 1 or 0 (on or off) — just like a memory cell. Electrostatic forces can then cause each mirror to either align so it transmits light to the TV screen (lighting up a pixel) or to move out of alignment (darkening a pixel).

Researchers are busy developing all kinds of MEMS that can manipulate tiny objects. Figure 8-4 shows a MEMS transmission (basically a bunch of gears) that takes the output of a microengine on a chip and then manipulates another object on the same chip.

Figure 8-4:
MEMS
transmission
to convert
microengine
output into
linear
motion.

Courtesy of Sandia National Laboratories, SUMMiTTM Technologies, www.mems.sandia.gov

In Figure 8-5, you can see that the output of the MEMS transmission is connected to a hinged mirror. The microengine is capable of adjusting the angle of the mirror.

Building computer brains from molecules

All we need now are some tiny brains (no cheap remarks, please) to go with the tiny fingers . . . and we're partway there already: Computer chips have been miraculously shrinking over the years to become more powerful. But as the components of chips — stuff like transistors, diodes, and memory cells — continue to shrink, there's less distance between those components; the increased density in the chip creates problems. Nanotechnology may take us the rest of the way to creating computer chips based on molecular-scale devices (or even smaller nano-devices) that can solve many of these problems.

Figure 8-5:
Hinged
mirror
connected
to the trans-
mission.

Courtesy of Sandia National Laboratories, SUMMiTTM Technologies, www.mems.sandia.gov

So, what's the problem?

Everybody wants small, powerful, fast computer chips — but making them smaller is a tricky business. A few of the biggest problems with shrinking computer chips are

- ✔ **Noise:** Leakage of electrons from one device — a transistor, for example — to another or from one part of a device to another. As the thickness of the insulating layers is reduced, their effectiveness is also reduced. This shrinkage increases leakage of electrons and produces electrical "noise" in the circuit.

- ✔ **Heat:** Generation of heat from having so many transistors and other devices in one small space.

- ✔ **Cost:** As the feature size of components shrinks, the cost of manufacturing equipment used to create the components increases.

How nano can help

Building transistors and other devices used in computer chips either from organic molecules (see the next section for more about these) or from nanowires, nanotubes, and nanoparticles the size of a molecule may be a way to minimize these problems. This technique is called *molecular electronics*. Nanotechnology may allow us to simply pack more onto each computer chip.

Here are some of the projected benefits of molecular electronics:

- ✔ **Quieter:** Reduced leakage of electrons (which happens because electrons in molecules are confined to certain orbitals and energy levels) means less electronic noise. Scientists believe that, even when a molecular transistor is turned on and current (that is, a stream of electrons) is flowing, the electrons travel within *overlapping* orbitals that end up controlling their movement and keeping them where they belong.

- ✔ **Cooler:** Reduced heat is one result of lowering the number of electrons used by the devices. Tens of thousands of electrons are currently used to turn a memory cell on and off (changing it from a 0 to a 1); but a molecular memory cell will probably need fewer than 100 electrons to change its state.

- ✔ **Cheaper:** Lower cost can result from the use of batching processes (such as those used in chemistry) to produce molecular computer chips. If successful, this technique could significantly reduce the cost of manufacturing equipment.

Using organic molecules

Organic molecules, also called *hydrocarbons,* make up the solid (as opposed to liquid) portions of living things (that is, people, animals, and plants), as well as certain materials such as plastic and oil. The number of carbon and hydrogen atoms bonded together into different shapes with other atoms (such as nitrogen or oxygen) can vary in organic molecules.

The difference between your dog Sparky and your plastic credit card is just the number of carbon and hydrogen atoms, the types of atoms that are tacked on, and the way the atoms are bonded together in the molecule. These factors, in combination, determine the properties of any organic molecule. (Okay, Sparky is cuter, too, and won't fit in your wallet.)

Two major requirements for using organic molecules in molecular computing are important for you to get your mind around:

- ✔ Electrons have to be able to flow easily through a molecule.
- ✔ You have to be able to turn the flow of electrons on and off.

Electrons flowing freely

In some organic molecules, the electrons are organized in orbitals in such a way that they can travel between the atoms in the molecule. These gadabouts are called *delocalized electrons* (like those that exist in metals). Benzene, for example — a molecule that's a ring of six carbon atoms (see Figure 8-6) — has delocalized electrons. Organic molecules that contain these delocalized electrons meet the first requirement of molecular electronics — they allow the free flow of electrons, God bless their little hearts.

Figure 8-6:
A benzene ring, illustrating the delocalized electrons.

Delocailzed electrons

Benzene
(C_6H_6)

Turning the flow on and off

A lamp you can't turn on and off isn't much use. Uncontrolled electrons are sort of the same way. The second requirement of molecular electronics is that you have to be able to turn the flow of electrons on and off. Fortunately, researchers have done that with organic molecules — in particular, a molecule in which three benzene rings are connected in a line; delocalized electrons can move within all three benzene rings. Change the center benzene ring a little (by adding nitrous dioxide to one side and ammonia to another side), and you skew the electron orbitals in the center benzene ring so they're asymmetrical — preventing the electrons from moving from one ring to another. In this way you can block (turn off) the flow of electrons.

By applying an electric field, you can untwist that benzene ring like a faucet handle and allow electrons to flow. What you've got here is an electrically operated switch like the one shown in Figure 8-7.

Okay, what about other tiny components? Hmm . . . tricky. To create (for example) a molecular transistor that's equivalent to the standard transistors shown in Figure 8-7, researchers have to determine how to turn the flow of electrons on and off by applying a voltage directly to part of the molecule, rather than applying an electric field indirectly to the whole thing.

Adding things to molecules

One way to change an ordinary molecule into a component of a tiny machine is to add a specific number of specific atoms to it — just here, just there, just so. Adding parts to a molecule is essentially what chemists do when they make new materials. Certain molecules in a chemical reaction exchange atoms or groups of atoms with certain other molecules to produce new molecules. The chemist "merely" figures out what molecules to mix together under what conditions to produce a new molecule, and then devises procedures to separate the molecule of interest from the entire brew. Chemical reactions are done as batch processes. If we figure out how to create molecular devices, we'll be using batch chemical processes to create them as well.

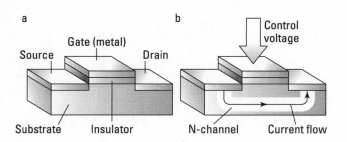

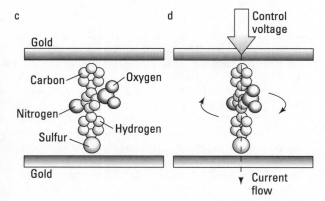

Figure 8-7:
Molecular switches compared to transistors.

Courtesy of Scientific American

Using nanotubes and nanowires

Nanotubes and nanowires are getting a lot of attention for a variety of electronic applications. Nanotubes are cylinders of carbon atoms; nanowires are tiny wires made from metals such as silver or semiconductors such as silicon.

Researchers have already created memory devices using nanotubes and nanowires. (See Chapter 6 for more about this.) Eventually the computer industry may be able to replace memory cells that are about a hundred nanometers wide with cells that are only a few nanometers wide. Hewlett-Packard (for example) recently found a way to use nanowires to construct a device — only a few nanometers wide — that performs electrical operations normally requiring at least two transistors. This device, a special combination of three nanowires, is called a *crossbar latch*. Crossbar latches can replace the transistors used today with nano-scale devices that do the work of transistors.

Do it yourself: Self-assembly

Of all the features of molecules that scientists are eager to emulate, one will probably do the most to make construction of a molecular computer easier: *self-assembly*. It's a pretty simple phenomenon: When molecules bond together, they just naturally determine their proper orientation. No fuss, no muss.

One example of how self-assembly could be used in the construction of a molecular computer looks like this: Suppose you attach a sulfur atom to one end of a molecule designed as a transistor or other computer component. It turns out (who knew?) that sulfur attaches itself to gold. If you place a solution containing a molecular device on gold-coated material, the sulfur end of each molecule in the device attaches itself to the gold surface. You end up with an array of these molecules sticking up from the gold surface like miniature blades of grass, as seen in Figure 8-8.

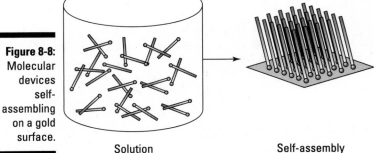

Figure 8-8: Molecular devices self-assembling on a gold surface.

Solution   Self-assembly

But the molecules don't plunk down at random on the gold surface — rather, they line up in regularly spaced rows and columns. Okay, assuming there's no miniscule drill sergeant down there ("All you sulfur slackers, ten-HUT!"), what gives?

Well, the answer's repulsive. Really. In Chapter 3 we talk about the fact that electrons repel each other — and that the repulsive force between them helps determine the shape of an atom. Following that line of thinking, when you put a gold-coated surface in a solution containing these molecules, they all rush toward the gold. As the molecules get closer to the surface, however, the electrons surrounding each molecule push each other away.

Researchers are investigating atoms other than sulfur that bind to different metal or semiconductor surfaces. Using this property of self-assembly should enable us to program each end of a molecular device to attach to a different type of surface. Result: components.

Here's another neat self-assembly trick. Let's say you've built several different types of molecular devices that you want to attach to various portions of a computer chip. You apply a voltage to a portion of the chip while it's immersed in a solution that contains one type of molecular device. The devices attach to the area of the chip where you're applying the voltage.

Anna Belcher is leading a team of researchers at the University of Texas to take another path to a self-assembly process, using some tricks from biology. They have found proteins that will bond to certain metals or semiconductors and are using this capability, along with the ability of proteins to assemble into structures such as an abalone shell, to develop a self-assembly process for computer chips.

Wire it up

Once we figure out how to make all these molecular transistors or memory cells, we still have one more problem to address: How do you connect these things together or connect them to an external circuit?

Nope, this is one time duct tape won't work. What's needed is a type of wire that has a diameter of no more than a few nanometers. You then have to be able to connect that wire to the molecular transistors or memory cells. (And if you've skimmed this book, you know better than to grab the tweezers to do this.)

Nanotubes and nanowires are likely candidates because they can be made only a few nanometers wide and can conduct electricity. Nanotubes can be either metallic or semiconductor-like — so one piece of this puzzle is to sort out the metallic nanotubes to use for this particular application.

Another candidate for connecting molecular transistors or memory cells is conductive organic molecules, called *molecular wires*. Molecular wires are strings of molecules whose atoms have orbitals that allow delocalized electrons. Studious types at Rice University have been looking into the procedures required to synthesize *molecular wires*. When you bond a bunch of conductive organic molecules together, you produce a much longer molecule — in which the orbitals allow electrons to travel the entire length of the molecule. Another way of saying this: The molecule conducts electricity.

Whether you're using a nanotube or molecular wire, sulfur (or some other atoms that bond to specific metals) is applied to each end to attach the molecule to metal surfaces or other devices (see the earlier section "Do it yourself: Self-assembly" to understand how this works).

There's still a lot of work ahead of us, however, before we find ways to *route* molecular-scale wires (whether nanotubes, nanowires, or organic molecules) to specific devices and then on to the next device required to complete the circuit. But the methods discussed in this chapter provide the groundwork for these cool applications of nanotechnology. Build it, and it (whatever nano device you have in mind) will come.

Chapter 9

Getting Energy and a Cleaner Environment with Nanotech

*T*he world is a populous place, and mankind has a tendency to use up energy on everything from heating our homes to driving to work. Just imagine if we had unlimited energy to fulfill the needs of these billions upon billions of people. With unlimited energy, we could effectively desalinate water from the oceans to irrigate crops, greatly expanding our food supply. We could get beyond our dependence on a shrinking supply of fossil fuels, whose use can harm our environment. Nanotechnology may well be the key that allows us to not only efficiently utilize our current resources, but also create additional sources of energy and help our struggling environment.

In this chapter, we take a look at ways we can use nanotechnology to produce energy with more cost-effective solar cells, more efficient hydrogen production, better storage possibilities for hydrogen fuel cells, and more efficient batteries. You get a look at using nanocatalysts to support chemical reactions, and glimpse how nanotechnology may improve our air and water supplies.

The Energy Challenge

A limitless supply of energy is our new Holy Grail. The future of our expanding, technology-based society depends on our finding that Grail. Don't believe us? Check out Figure 9-1, which illustrates how the projected demands on the world's energy resources will soon exceed the supply unless non-fossil-based sources of energy increase significantly.

The energy problem needs to be fixed, and soon — and nanotechnology may be the tool we can use to fix it. Using nanotechnology we can explore a variety of alternate sources of energy, from solar cells to hydrogen fuel cells to nano-efficient batteries.

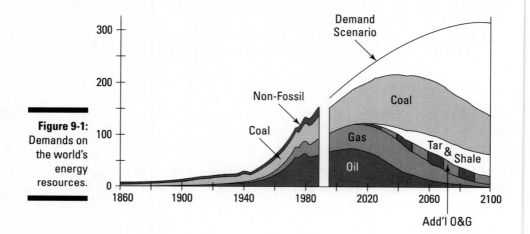

Figure 9-1:
Demands on the world's energy resources.

Using Nanotechnology to Make Solar Cells Affordable

When light hits a solar cell, an electric current is generated. When you punch numbers into a calculator without inserting a battery or plugging it into the wall, you're taking advantage of solar cells.

Today, people typically use solar cells in places where they can't use standard electrical outlets — say, in sailboats or satellites. Solar cells are also used in small machines like calculators where frequently replacing batteries is inconvenient and costly, and the amount of power required from the solar cells is so small that they aren't too expensive to produce.

Solar-cell sticker shock

Where the cost of solar cells gets prohibitive is in larger installations. Though there may be a family in your town using solar cells to power their home's heat or lights, this is still a pretty rare scenario. That's because it's not yet affordable to "go solar" in any big way. Manufacturing and installation costs are such that anyone actually using solar is probably doing so on principle or because they're off the grid.

But why is it so costly to make solar cells today? The answer to this one is pretty much a no-brainer: A solar cell is made with semiconductor material, which presently requires expensive equipment and high temperatures (several hundred degrees centigrade). This expensive manufacturing process drives up the cost of the solar cell. In fact, when you figure out the cost of electricity generated by the solar cell over its lifetime, you're looking at about 40 cents per kilowatt hour. This cost is several times the rate your power company is charging you (rates vary depending on where you live, but for one of us, a recent bill came in at under 8 cents per kilowatt hour — which isn't unusual).

The potential of nano solar cells

Various companies are using nanotechnology to reduce the cost of making solar cells to the point where solar cells can compete with . . . the power you buy from your local utility company. These nano-style solar cells use semi-conducting nanotubes, nanowires, or nanoparticles embedded in a conductive plastic. These cells work just like the solar cells currently available — but they cost much less to manufacture.

For example, a company called Nanosolar, Inc., is developing solar cells using a method that sprays or prints the layers of a solar cell onto a surface, much the way an ink-jet printer sprays ink onto a page. Nanosolar claims that not only is this process lower in cost than current solar cell manufacturing methods, but the resulting solar cells will be considerably thinner and lighter. The lighter weight will make it easier to cover an entire roof with solar cells.

Another company, Konarka Technology, Inc., is developing solar cells that use titanium oxide nanocrystals embedded in plastic. These cells can be used in devices like laptops and cellphones.

Konarka has also demonstrated a process to make fabrics that work as solar cells. The solar cells that you can fit into a jacket in the near future will probably provide a small trickle of electricity, about enough to keep the battery for your cellphone charged. But consider other possibilities: With such a fabric you could make a tent that could provide a convenient light source when you're camping in the woods.

Another application suggested by Nanosys, Inc., involves installing solar cells between the glass panes of windows. Solar cells for this application would probably have to be transparent. Get this approach to work, and all the windows in a giant skyscraper could power every office in the building.

How, exactly, do nano solar cells get built?

Quite a few companies are racing to get their share of the marketplace for less expensive solar cells, but how will they get there?

Different companies are pursuing different nanotechnological approaches to developing solar cells, but the general idea is the same for all. When light hits an atom in a semiconductor, those photons of light with lots of energy can push an electron out of its nice stable orbital around the atom. The electron is then free to move from atom to atom, like the electrons in a piece of metal when it conducts electricity.

Using nano-size bits of semiconductor (we discuss semiconductors in Chapter 3) embedded in a conductive plastic maximizes the chance that an electron can escape the nanoparticle and reach the conductive plastic before it is "trapped" by another atom that has also been stripped of an electron. Once in the plastic, the electron can travel happily through wires connecting the solar cell to your gadget (cellphone, laptop, or whatever). It can then wander back to the nanocrystal to join an atom that has a positive charge. (For the semiconductor physicists among you, this is called *electron hole recombination*.)

Figure 9-2 illustrates the workings of a solar cell that uses nanotechnology.

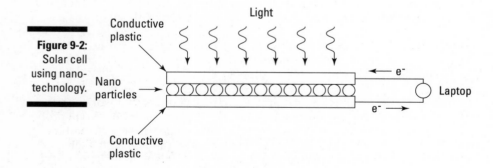

Figure 9-2: Solar cell using nanotechnology.

Making Hydrogen Fuel Cells

Here's a power source, long used in spacecraft, brought down to earth: A *hydrogen fuel cell* takes oxygen and hydrogen and generates electricity and water. Hydrogen fuel cells could potentially prove to be a great power source because they don't pollute the air. But current methods of producing hydrogen do pollute the air. Another challenge with hydrogen fuel cells is the problem of hydrogen storage. If we could only store sufficient volumes of hydrogen in a tank in a car, for example, we could power cars more cleanly.

Before we get to efficient production and use of hydrogen fuel cells, we have to solve these problems.

It's a matter of density

If you have a supply of liquefied dinosaur remains handy (petroleum, to you), you'll be glad to know that this material has a major advantage over many electric and mechanical storage systems. Petroleum, from which we also produce gasoline, has serious *energy density* — it generates a lot of power from a compact volume of material.

For example, the amount of energy stored in a gallon of gasoline, pound for pound, is more than six times the amount of energy stored in most batteries. When designing a car, it's a good idea to minimize the weight and volume of fuel required. That's why car designers are hesitant to switch to electric cars: A tank of gas weighs only a fraction of the weight of the batteries you need to take a car the same distance. On top of that, you can fill up your car with gas in minutes; most batteries take hours to charge.

Of all the alternatives to fossil fuels, hydrogen gas provides the soundest solution. Hydrogen gas combusts to form water — a benign environmental element — and its energy density is reasonably close to that of petroleum. When you add that it can be derived from salt water — one of the most plentiful resources on the planet — it becomes a very appealing option.

Putting hydrogen into production

We're going to need nanotechnology to improve two major components of the hydrogen fuel-cell system if we're ever going to make fuel cells usable in our cars. The first component that needs major improvement involves actually producing the hydrogen with which to fuel the cells.

The method most often used to produce hydrogen today is called *steam reforming*. This method uses natural gas — and generates air pollution in the form of carbon dioxide. Another method called *electrolysis* uses electricity to produce hydrogen from water. Electrolysis (not to be confused with hair removal) does not produce any air pollution in itself, but you have to get the electricity from somewhere — generally from coal, oil, or natural-gas power plants that *do* generate air pollution.

Nanotechnology researchers are trying to develop a cleaner method of producing hydrogen. Curious? Read on to get the full story.

If you really want to know . . .

There are other fuels besides gasoline with high energy densities — some right under your nose. Animal fat is very dense in energy, and has been used to power some unique prototype vehicles. Even a combination of urine and pickling vinegar can get you from Point A to Point B if you're desperate enough (after all, somebody *was* desperate enough to invent the method). But there's no reason to get that desperate: Just wait ten or so years and nano may bring you a far less odorous solution.

Producing hydrogen with photoelectrochemical cells

One method of producing hydrogen has been tagged with a lengthy name: *photoelectrochemical hydrogen production*.

The working part of a photoelectrochemical-hydrogen-production device is a layer of nanoparticles, roughly 30nm in diameter, placed on a conducting glass layer, as shown in Figure 9-3. The nanoparticles are composed of a semi-conducting metal oxide (say, an oxide formed of titanium). The conducting glass plate is connected to a metal electrode located a small distance from the layer of nanoparticles. You fill the space between the electrode and nanoparticles with water. Water seeps into the space between the nanoparticles, providing the most possible surface contact between the semiconducting metal oxide and the water. (Read the nearby sidebar for more about why this is important.)

To get a handle on how a photoelectrochemical cell works, think for a moment about the way electrons move: Light strikes a layer of nanoparticles, knocking electrons loose from some atoms — just like what happens inside a solar cell. Those electrons move through the conducting glass layer to the metal electrode and place a negative charge on it.

So now you have a layer of nanoparticles that electrons are rushing out of — and a nearby metal electrode that electrons are rushing into. Water filling the space between the two interacts with the electrons, producing hydrogen.

In practice, the photoelectrochemical cell needs a little help to make a reaction occur. By attaching a solar cell, you can provide the extra energy required to the electrons in the photoelectrochemical cell.

By this process, a photoelectrochemical cell makes hydrogen from water and sunlight. Various researchers believe that this is the near-ideal model for replacing the current method of using natural gas to produce hydrogen. In fact, one company, Hydrogen Solar, Ltd., is building an experimental power plant in Las Vegas, Nevada, to demonstrate just this process.

Producing hydrogen: Behind the scenes

If you have a truly scientific bent, you may want some more details about the Photoelectrochemical Cell method of producing hydrogen. Water happens to be composed of two hydrogen atoms and one oxygen atom. (Remember H_2O from Chemistry 101?) The only reason these atoms stay together in a molecule is because they have this primal urge to share electrons. In photoelectrochemical cells, some of the water molecules touching the metal-oxide nanoparticles give up electrons to replace those knocked out of the metal-oxide atoms by the photons. Rather than just resting on the surface of the metal oxide, water fills the entire space between the nanoparticles — making more water molecules available to give up electrons, which speeds up the photoelectrochemical process.

While water molecules are giving up electrons to the metal-oxide nanoparticles, the water molecule breaks up into atoms. Because hydrogen atoms have a weaker hold on electrons than the oxygen atoms, you end up with hydrogen atoms that have no electrons — *ions* — in which their remaining protons give them a positive charge. Because unlike charges attract, the negatively charged metal electrode pulls these hydrogen ions over to it. When the hydrogen ions reach the electrode, each one grabs an electron and becomes a neutral hydrogen atom. Then, two by two, these atoms combine to produce molecules of hydrogen. As a byproduct, this process also produces oxygen.

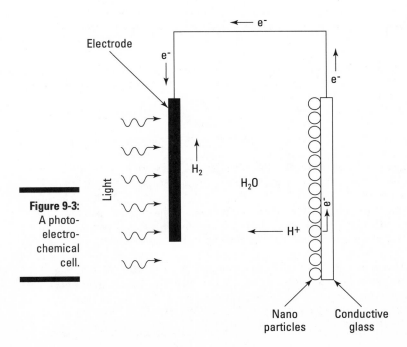

Figure 9-3:
A photoelectrochemical cell.

Electrode

e^-

e^-

e^-

Light

H_2

H_2O

e^-

H^+

Nano particles

Conductive glass

Producing hydrogen with designer molecules

Plants — the biological kind — are pretty smart creatures in the way they work. A plant takes in sunlight, carbon dioxide, and water, and releases oxygen. Scientists have been looking at the process that plants use to produce oxygen to see whether we can emulate it.

So what happens inside the plant? When a molecule called chlorophyll absorbs sunlight, electrons in the chlorophyll are pushed from their orbitals. The electrons, water, and carbon dioxide react to form sugar through the process of *photosynthesis.* The plant uses this sugar as food — and produces, as a byproduct, the oxygen we breathe every day. At the end of this process, the chlorophyll molecule is unchanged — it can go back to help change other water and carbon dioxide molecules into sugar and oxygen.

A molecule like chlorophyll that assists in a reaction but is unchanged by it and hangs around to assist other molecules in the reaction is called a *catalyst.*

Researchers are attempting to design molecules that absorb sunlight and produce hydrogen from water, just as chlorophyll produces oxygen from water. At Virginia Tech, some folks have noted that molecules containing atoms of ruthenium can absorb light and produce electrons, so they put an atom of ruthenium at each end of a designer molecule. They also put an atom of rhodium in the center of the molecule to transfer electrons to water. They filled out the structure of their designer molecule with atoms of carbon, hydrogen, chlorine, and nitrogen — and came up with a molecule that produces hydrogen from sunlight and water, just as chlorophyll produces oxygen.

Although still in the laboratory stages, this research holds out the promise of using a clean process — similar to one in nature — to produce hydrogen for generating energy.

Who are ruthenium and rhodium when they're at home?

What are *ruthenium* and *rhodium* and how do researchers put this designer molecule together? Ruthenium and rhodium are two singing sisters who . . . just kidding . . . actually they're both metals that are similar to platinum. One reason you may not have heard of them is that they are so scarce that they cost more to obtain than gold or platinum. Researchers put this designer molecule together by causing a chemical reaction between different molecules that they want to combine. But this is not something for you to try at home: It typically takes researchers several experiments and sometimes years of work to figure out how to build a designer molecule.

It starts out efficient, but then . . .

Researchers at the Max Planck Institute are trying to improve the ability of existing materials to store hydrogen. They added titanium nanoparticles to a substance called sodium aluminum hydride — and improved its ability to absorb and store hydrogen. The modified material absorbs 4.5 percent of hydrogen, about 40 times faster than sodium metal hydride itself. Refilling this material with hydrogen would take about 15 minutes. There are still issues to be resolved; for example, after 17 cycles of filling and emptying the material of hydrogen, fill-up time slows way down to about 45 minutes.

Storing hydrogen

Hydrogen fuel cells pose one other major problem to nanotechnology researchers: After we've generated hydrogen, exactly how do we *store* it? If hydrogen fuel cells are to be effective in automobiles, hydrogen storage systems should store 6 percent (by weight) of hydrogen compared to the weight of the storage tank. In addition, you should be able to remove the hydrogen easily when you need it. Now, conventional methods store hydrogen as a compressed gas or as a liquid. To store enough of it to power a car (in these forms, anyway), you'd need a tank the size of Rhode Island. (Okay, maybe not that big, but it would definitely be too heavy and too large to put in your Chevy.)

Researchers are attempting to develop lightweight, compact solid materials that do the same job as a storage tank: These materials would absorb hydrogen the way a sponge absorbs water, and hold it (without leaking) until you need it. Theoretically, these materials could be refilled with hydrogen in a reasonable amount of time (perhaps about ten minutes) and be reused many times.

Although materials exist that currently perform this role, they only store about 1 percent of hydrogen by weight (versus the 6-percent-by-weight required). Researchers are looking at using nanoparticles to enhance the properties of existing solids — as well as the possibility of developing entirely new materials.

One such new material that has been proposed to store hydrogen is made of carbon nanotubes. Experimental results at this point are mixed; the reasons for these differences aren't clear. For example, researchers at the National Renewable Energy Laboratory have reported that carbon nanotubes, in the presence of certain metals, can hold 8 percent by weight hydrogen; other results report achieving only 1 percent. Work will continue to optimize the ability of porous materials that contain nanotubes to store hydrogen — and to understand why these experiments get different results (so far).

Using Nanotechnology to Energize Batteries

Of all batteries currently in use, the *lithium-ion* type stores the most electrical power for its weight. The bad news is that (for now, at least) you can only use this type of battery in things like watches and laptop computers. These are devices that do not have sudden demands for a lot of power, as do power tools. No surprise that researchers are using nanotechnology to improve lithium-ion batteries so they can be used in more devices.

When a battery runs your radio or other gadget, the reaction of chemicals in the battery transfers electrons to the *anode* — a piece of metal that makes up the negative terminal of the battery. Those electrons become the electric current that powers your gadget. One way to improve batteries is to make the anode out of material that lets you maximize its surface area.

In a lithium-ion battery, the anode is currently made of carbon. Altair Nanotechnologies, Inc., is developing an improved lithium-ion battery that replaces the carbon anode with one made up of lithium titanate nanocrystals. An anode made up of these nanocrystals provides about 30 times greater surface area than one made up of carbon. If the anode has greater surface area, electrons come out of the battery faster; this means that a higher electric current — therefore more power — is available to run your gadget.

If electrons can leave a battery faster, they can also go back in faster, trimming the time it takes to recharge the battery. Assume it takes an hour to recharge your battery-powered drill, and that you have to get a new battery after only 500 charges. Altair Nanotechnologies projects that you can recharge their nano-batteries in a few minutes, and the batteries will last for several thousand recharges.

Toshiba's researchers announced, as we're writing this chapter, that they have used nanoparticles in developing a lithium-ion battery that recharges to 80 percent of its capacity in one minute. Toshiba stated that a prototype of this battery has been discharged and recharged 1000 times with only a 1-percent decrease in its capacity. The only details Toshiba has released is that nanoparticles are used in the anode.

Such developments should allow us to replace current batteries (not only lightweight lithium-ion batteries, but also more powerful batteries such as nickel-metal-hydride ones) with lighter-weight models that recharge faster. For those of you who travel with laptops, you'll be happy to hear that this should significantly reduce the amount of time your battery needs to recharge before you head off to that next (inevitable) meeting.

It just keeps on going

Another company, mPhase Technologies, is using carbon nanotubes to develop batteries with a very long shelf life. A normal battery, even when not supplying current, has some reactions going on at a low level between its internal chemicals and its electrodes — and those reduce its ability to deliver electric current when you need it. That's why mPhase is developing a battery that contains nanotubes attached to the surface of the electrodes. The chemicals (in liquid form in this case) rest on the top of the nanotubes, which are held in place by surface tension. Because the nanotubes keep the chemicals separate from the electrodes, no reaction takes place — so there's no degradation of battery power. When you need current from the battery, you apply a voltage, which causes the liquid chemical to drop between the nanotubes and reach the electrodes: Voilà! You've got power!!

Using Nanotechnology to Reduce Energy Consumption

It's sad but true: Much of the energy produced in the world is wasted. For example, a light bulb may produce light, but it also produces heat — which may be fine in winter but a total waste on a hot August day. Other processes take more energy than may be necessary to make them happen. Making gasoline, for example, requires applying heat to petroleum to cause a reaction. Finding ways to save energy in these processes is important to reducing our energy consumption. The following sections highlight some of the more promising nano-paths researchers are taking in trying to keep energy consumption down.

Producing light with nanotechnology

Actually only about 5 or 10 percent of the power from an electric current running through an incandescent light bulb generates light; about 90 percent is spent generating heat. Researchers are projecting that light bulbs made with quantum dots (nanocrystals that emit visible light when exposed to ultraviolet light, as described in Chapter 8) will turn *almost 100 percent* of the power from electricity into light. With quantum dot bulbs, very little energy will be wasted as unneeded heat.

Given that about 20 percent of the electric power consumed in the world is used to generate light, adopting light bulbs based on quantum dots could cause significant reduction in overall energy consumption *worldwide*.

Using nanocatalysts to make chemicals

Another way in which nanotechnology has an opportunity to reduce energy consumption is in the production of chemicals. *Catalysts* have been traditionally used to reduce the amount of energy required during chemical reactions. Using nanotechnology, we can create even more efficient catalysts and save even more energy.

Using catalysts to use less energy

Often, when molecules are formed in a chemical reaction, energy — usually in the form of heat — is needed to break apart the atoms in the incoming molecules and rearrange them into new molecules. The brute-force method is to add enough heat so the molecules smash into each other, breaking off part of one molecule so it can bond to another molecule. This reaction forms the new molecule you want to produce.

Chemists have long used catalysts to reduce the amount of energy needed to make a reaction work. An example of a traditional catalyst that can be used in this way is the platinum you find in the catalytic converter in your car. The platinum assists in the reaction that converts poisonous carbon monoxide and nitric oxide to less-harmful carbon dioxide and nitrogen. Here's how this works: When nitric oxide lands on platinum, the nitrogen atoms and the oxygen atoms bond to the platinum atoms, breaking up the nitric oxide molecule. The nitrogen atoms are then free to bond with other nitrogen atoms and are released as nitrogen gas. The oxygen atoms bond with carbon monoxide molecules that have attached to the surface of the platinum — and the result is carbon dioxide, also released as a gas. The platinum surface is available for other nitric oxide molecules to land on, and the process continues. By using the natural tendency of the molecules to bond to the platinum rather than using heat to break up the molecules, you reduce the amount of energy required for the chemical reaction to occur. Also, the fact that the catalyst can be used over and over again means you don't have to replace the material.

Traditional catalysts do their jobs very nicely, thank you very much, but in life there is always room for improvement. Making the jump to nanocatalysts would allow companies to produce complex chemicals — such as the gasoline produced from crude oil — using way less energy. Read on to see how this might work.

Here's where nanocatalysts come in

A catalyst with a large surface area can get more done because it interacts with more molecules at the same time. Using nanoparticles or nanocrystals to form catalysts provides larger surface area materials — so researchers are looking to nanotechnology to improve the performance of existing catalysts.

In fact, many existing catalysts contain nano-size particles; not until recently, however, did scientists have the tools or knowledge to study particles this small. Now researchers are designing and developing nanoparticle catalysts with a greater understanding of how these materials work on the nano scale.

At one end of the ready-for-nano-prime-time spectrum is NanoGram Corporation, which has developed — and is already marketing — catalysts composed of nanoparticles. Their catalysts can offer improved purity and size uniformity compared to previously available catalysts. They claim this will improve the effectiveness of the catalysts at a relatively low cost.

Scientists at Georgia Tech, however, are still focused on the research phase, struggling to gain even more knowledge about how catalysts work. These researchers have found that the ability of gold nanoparticles to act as a catalyst depends upon the condition of the substrate the nanoparticles rest on. They looked at how nanoparticles containing eight gold atoms on a magnesium oxide substrate interact with carbon monoxide. It turns out that if a gold nanoparticle is attached to a substrate where there is a defect in the magnesium oxide film, the nanoparticle transfers an electron to the carbon monoxide. This does not occur if the gold nanocluster is attached to the substrate at a location without a defect.

The researchers at Georgia Tech are looking farther into how the condition of the substrate and the size of the nanoparticle affect the efficiency of the catalyst; they hope it may lead to catalysts that allow reactions to occur at lower temperatures. A successful result would require still less energy to produce a reaction.

How Nanotechnology Can Help Our Environment

One of the big challenges in producing energy is to do it in a way that keeps our planet from getting filthy. If nano-based solar cells succeed in producing electricity, it will be at a lower cost than power plants that burn hydrocarbons such as oil, natural gas, or coal. If our world reduces the use of such power plants as a result, we will greatly reduce associated air pollution.

If we can make the production and storage of hydrogen effective and cheap, we may see widespread use of hydrogen fuel cells in cars, trucks, and ships, which will also help our planet stay cleaner.

We hope that, in the long term, nanotechnologists will succeed in these areas, but there are other, perhaps more incremental, ways that nanotechnology can help our environment today.

Clearing the air with nanotechnology

One of our global problems is the increasing amount of carbon dioxide in the air. Instead of waiting around for cheap solar power to eliminate those hydrocarbon-burning power plants at some distant point in the future, we could try to clean up what's coming out of those smokestacks right now. Nanotechnology may be able to provide methods for doing this that are more effective and less expensive than methods we use now.

Researchers at Oak Ridge National Laboratory have developed a nanocrystal that might just do the trick. When carbon dioxide lands on a nanocrystal composed of cadmium, selenium, and indium, the nanocrystal donates an electron to the carbon dioxide. This extra electron allows the carbon dioxide to react with other molecules in the smokestack — and become harmless. In effect, the nanocrystal acts as a catalyst. If filters containing these nanocrystals can be made cheaply enough, putting them in smokestacks can significantly reduce emissions of carbon dioxide.

Another nasty emission that researchers are hoping nanocrystals can help us deal with is mercury vapor. Mercury vapor is emitted by coal-fired power plants. One mercury-containment method being investigated uses titanium oxide nanocrystals under UV light — which turns the mercury vapor into mercury oxide, a solid. However, placing a UV light source in the smokestack poses a possible maintenance issue. Instead, we might use nanocrystals of iron oxide (essentially high-tech rust), which require heat rather than UV light to make the reaction work.

Then there are the diesel engines that emit nitrogen oxides, which you've probably breathed if you've been stuck in traffic with big rigs and buses. Biofriendly Corporation has been working on it; they've developed a nanocatalyst that, when added to diesel fuel, causes it to burn up more completely. That means fewer nitrogen oxides are emitted. Biofriendly has received two grants from the State of Texas to demonstate this nanocatalyst. (The second grant is just to help them work through the Enviromental Protection Agency verification process, which may prove harder than solving the original technical problem.)

Don't think that "clean" manufacturing industries — those industries that aren't typically thought of as polluting the enviroment with waste materials or toxic emissions, such as computer chip manufacturing — are worry-free. It turns out that many of these industries use organic chemicals in their manufacturing processes. Organic chemical vapors can themselves be harmful. Researchers at the Pacific Northwest National Lab, among others, are investigating nanomaterials that we could use in filters to trap organic chemical vapors escaping from such factories.

Keeping water crystal-clear with nanotechnology

Life needs water, but many of our lakes and streams have been contaminated by wastes from industrial plants — add to that the pesticides used in our gardens or by farms and you've got a serious problem. While current laws have reduced the amount of contamination going into our waters, there are still lakes and streams that are significantly contaminated. Researchers are looking at ways that nanomaterials can help to clean up our water act.

Getting rid of TCE

One example is a joint venture: Rice University and Georgia Institute of Technology are developing a better way to remove TCE (trichloroethylene) from water. TCE has been found at a majority of sites on the EPA's Superfund list, and it's pretty horrible stuff that can cause heart problems, nausea, vomiting, and eye irritation. TCE is primarily used to degrease components during manufacturing operations but it is also used in products such as spot removers for clothing (cleanliness comes at a price).

Palladium acts as a catalyst to convert TCE to ethane, which is not horrible. But there's a problem: Palladium is a rare metal, more expensive than gold. Scientists are in search of the most efficient way to use palladium to neutralize TCE. Given the large number of sites that must be decontaminated, it's important to find a way to use as little of this expensive metal as possible.

Both nanoparticles made of palladium and nanoparticles made of gold coated with a layer of palladium atoms are possible options. Coating gold nanoparticles with palladium atoms makes all the palladium atoms available to catalyze the TCE molecules, and seems to be the most cost-effective use of palladium.

One way to get nanoparticles into the contaminated groundwater to do their work is to place a filter containing the nanoparticles in a pump that is used to circulate contaminated water. As the water passes through the pump, the TCE is broken down.

Making water fresh with solar energy

Distillation of salt water to make fresh water simply involves vaporizing the water to separate it from salt. If nano-based solar cells can produce cheap electricity, we might be able to use that electricity to turn salt water into water that can be drunk and used for agriculture in deserts. This could be useful in places like Los Angeles that now consume water gathered from watersheds hundreds of miles away. Countries with hot, dry climates that don't have enough water to irrigate much of their land (Egypt, for example) could also benefit. With enough cheap electricity, distillation could become economically feasible.

Cleaner water for less money

One direction being looked at for cleaning water involves production of nanoparticles that use less expensive materials than palladium, the current industry standard. One idea: Inject iron nanoparticles into a contaminated body of water, as illustrated in Figure 9-4. The particles would then spread throughout the water, cleaning it in the process. This would be much faster and cheaper than conventional methods that involve pumping the contaminated water out of the ground before treating it.

The iron nanoparticles are small enough that water could carry them into the soil that the water is contaminating. When they react with oxygen in the water, the iron nanoparticles turn into rust — a reaction that helps neutralize contaminants it comes into contact with.

How, exactly, does rust neutralize contaminants? When iron rusts, it reacts with both water (which contains an oxygen atom and two hydrogen atoms) and oxygen gas within the water. During this reaction, parts of molecules are separated and then recombined to form another molecule. As a result, some molecules that have either more or less electrons than normal (called ions) are present in the water that surrounds the iron nanoparticle. These ions are very reactive; when they encounter one of the contaminate molecules in the water, they react with it, producing a new, hopefully less dangerous molecule.

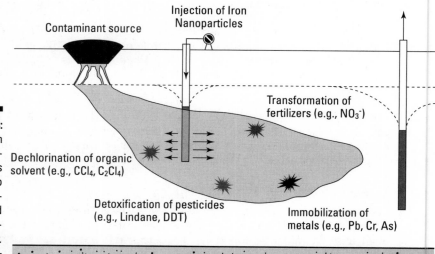

Figure 9-4: Iron nanoparticles injected into contaminated groundwater.

Contaminant source

Injection of Iron Nanoparticles

Transformation of fertilizers (e.g., NO_3^-)

Dechlorination of organic solvent (e.g., CCl_4, C_2Cl_4)

Detoxification of pesticides (e.g., Lindane, DDT)

Immobilization of metals (e.g., Pb, Cr, As)

Picking just the right iron nanoparticles

Rust? Useful? Who knew? Some scientists talk as if iron nanoparticles were the low-cost cure-all for groundwater contamination. The quality and effectiveness of iron nanoparticles, however, may vary depending on the lab that makes them. Different batches of iron nanoparticles, made with different processes, may give you different byproducts from their interaction with certain contaminants.

For example, one study showed that, in combination with carbon tetrachloride (toxic stuff used in many cleaning solutions), nanoparticles coated with an iron oxide containing boron turn carbon tetrachloride into chloroform — which is poisonous. Nanoparticles coated with an iron oxide containing sulfur turn carbon tetrachloride into harmless molecules.

While the idea of using iron nanoparticles may have great potential as a relatively low-cost way to decontaminate groundwater, it's really important to get the chemistry right: You have to consider the process that created the nanoparticle when you match it with a contaminant.

Part IV
Living Healthier Lives

The 5th Wave   By Rich Tennant

NANOTECHNOLOGY IN COMPUTER DEVELOPMENT

"Just don't sneeze again until we locate the servers, storage device, and mainframes."

In this part...

Could nanotechnology extend our lives by 10, 20, even 30 years? Who knows? But tweaking molecules and atoms — amazing to begin with — can accomplish even more amazing things. Some sober speculation suggests that what we learn about nanotechnology today could have a tremendous impact on health care (and therefore our longevity) in the future.

This part looks into some nano-applications that can improve our understanding of basic health — for instance, diagnosing medical conditions and mapping our genes. We also explore the role of nanomaterials in faster, better drug delivery, and in doing some tiny (but highly practical) chores — attacking cancer and other evil cells, combating heart disease, and knitting up broken bones.

Chapter 10

Diagnosing Personal Health Quickly, Easily, and Pain-Free

In This Chapter

▶ Utilizing microfluidics to diagnose a patient's problems

▶ Improving medical imaging through nanotechnology

▶ Mapping genes to cater our medicine

> *Dr. McCoy: "Well, your blood pressure is off the scale, if you call that green stuff in your veins blood."*
>
> *Spock: "The readings are perfectly normal for me, Doctor, and as for my anatomy being different from yours . . . I am delighted."*
>
> Star Trek *(1966)*

In the fictional 23rd century, Dr. McCoy diagnoses patients in an instant by waving a wand (okay, a "medical tricorder") over a patient's body. But even back here in the 21st century, diagnosis is going to start looking a lot like science-fiction mumbo jumbo. Imagine (for example) a patient's blood being analyzed in a lab the size of a computer chip, nano-size tracers racing through the body to light up tumors like Christmas lights, and our DNA being quickly mapped for us so we can get medical treatment customized down to tiny details. Before long, many of our current medical practices will look like the fumblings of the Dark Ages.

So let there be light! With nanotechnology allowing us to image our cells as easily as if we were taking a picture — even control them as if we were modeling clay — we are set to enter a completely new age of medical technology. Take a deep breath and join us as we make our way to the hospital to witness how nanotechnology will change the way doctors diagnose our ailments.

Lab-on-a-Chip

Current laboratories cost a lot in terms of space, time, and resources. Imagine if we could cut out sending samples to labs, push reactions faster, automate experiments, and do it all with a minute amount of precious samples and materials. This is the concept behind the *lab-on-a-chip* — miniaturizing entire labs into the space of a microchip. The vision here is a tool that could not only analyze our blood for viruses, but also detect chemical agents that terrorists could have released into our air and water — all in real time, allowing for a response in real time.

Dr. McCoy would certainly appreciate the space-saving nature of a lab-on-a-chip, considering the fact that space is very limited on vessels traveling in . . . space, that final frontier. The lab-on-a-chip could be used for experiments done in orbit or to take care of astronauts on the shuttle where laboratory equipment is too big and heavy to lug up. The applications are limitless, ranging from the family doctor's office to defense to space exploration.

Getting down to brass tacks for now, how does a lab-on-a-chip actually get its work done? In a traditional medical lab — the kind you see on all those hospital or CSI shows on television — you're going to see people wearing masks, vials of fluids, and large-scale scientific instruments. In lieu of all this, we shrink it down to the *size* of a computer chip but without all the electronic connections. Instead, we use micro- and nano-size channels to move and mix fluids — either liquids or gases — at the nano level. (Electronic circuits can be combined with our lab-on-a-chip but are not necessary and depend upon the application.)

Having trouble visualizing how this looks? Check out Figure 10-1, which shows a rendering of Agilent Technologies' lab-on-a-chip. Agilent is a spin-off of Hewlett-Packard, those masters of microfluidics (say what?) who have been producing finely blended inks for years for their ink-jet printers. (Although Agilent isn't the sole researcher or provider of lab-on-a-chip products, it's the one that caught our attention.) *Microfluidics* is the study of the behavior of fluids at volumes thousands of times smaller than a common droplet.

In this chapter, we talk a lot about microfluidics because we're convinced that the informational tidbits you glean from our microfluidics' discussion will help when it comes to getting your mind around fluid flow at the nano-realm.

As you might imagine, we run into a few problems when working with fluids at the micro and nano level. How do we fabricate channels through which we can pump our fluids? How do we move and mix those fluids? How do we analyze samples? In this chapter, we answer these and other fascinating questions by looking more closely at how these channels get made, how fluids are moved, and how samples are analyzed in your standard-issue lab-on-a-chip. As an added bonus, we also discuss how nanowires and carbon nanotubes can be used as a biosensor.

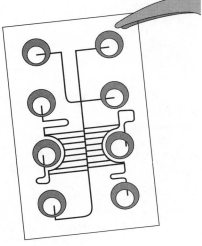

Figure 10-1: A lab-on-a-chip. Notice the micro-size channels, which move and mix fluids at the nano level.

Fabrication through soft lithography

Creating a lab-on-a-chip follows some of the same lithography techniques used to create the computer processor we discuss in Chapter 6. The objective is to produce nano-scale, leak-tight channels that allow for your fluids to get from Point A to Point B. In the world of processors we describe in Chapter 6, the goal is to build transistors layer-by-layer, one on top of the other. In this brand-new world of labs-on-a-chip, we want a layer, a channel, and another layer on top — kind of like a well-made sandwich.

In reaching our well-made-sandwich goal, we have quite a few fabrication options — as well as a number of choices when it comes to materials. Some fabrication methods involve etching (a method presently used for computer processors); another method involves a thick photoresist that is used to create tall, thin channels. (Photoresists use light-sensitive materials and chemicals to etch metals while layering our transistors. For more on photoresists, see Chapter 6.) One fabrication method we are particularly fond of is soft lithography, which uses polymers to create the nano-size channels of our lab-on-a-chip. Using polymers here is great because it allows for rapid prototyping and flexibility — not only in design but physically as well (that's what makes it "soft"). The upcoming steps take a quick walk through this particular fabrication method (pictured in Figure 10-2):

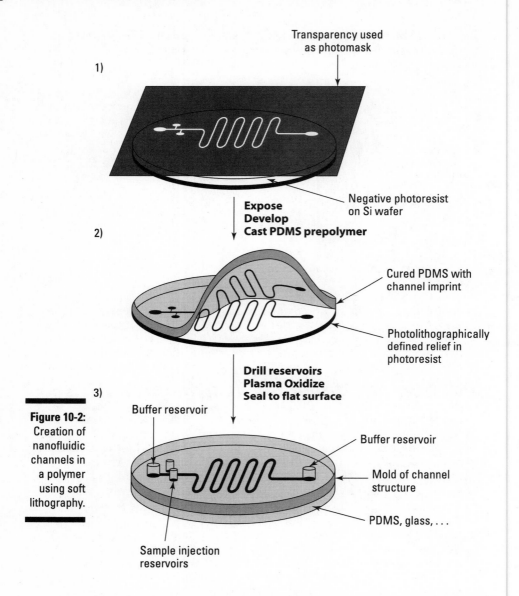

1)

Transparency used
as photomask

Expose
Develop
Cast PDMS prepolymer

2)

Negative photoresist
on Si wafer

Cured PDMS with
channel imprint

Photolithographically
defined relief in
photoresist

Drill reservoirs
Plasma Oxidize
Seal to flat surface

3)

Figure 10-2:
Creation of
nanofluidic
channels in
a polymer
using soft
lithography.

Buffer reservoir

Buffer reservoir

Mold of channel
structure

PDMS, glass, . . .

Sample injection
reservoirs

1. A master is created on a silicon wafer, using traditional processor-fabrication methods. To get a raised pattern that will act as a mold for our polymer, we use a *negative* photoresist: We etch away the surrounding material till what's left is a raised pattern that outlines the desired channels on the silicon wafer.

2. We then cure a polymer onto our silicon wafer mold, which now has a channel imprint with nanometer-scale features. *Curing* involves adding heat to make the liquid polymer harden a bit — ultimately it feels and

looks like Jell-O. This is the soft-lithography process. The polymer of choice is poly(dimethylsiloxane), PDMS for short.

3. We peel the cured layer of polymer off our master.

4. We oxidize the layer of polymer we just peeled off, sealing it to another layer — usually PDMS, silicon, or glass. This oxidation process is carried out by exposing the surface to plasma, a gas of charged particles, converting the surface molecules to its oxide by combining it with oxygen. If this oxidized surface (Si-OH) came in contact with another silicon-based (Si) material (like another PDMS, silicon wafer, or glass), the process forms an irreversible covalent O-Si-O bond.

5. We then add holes (an injection reservoir) to the polymer, allowing us to inject liquids. Once the fluid is injected, it flows through the nanochannels either mixing or separating molecules — all the work of a lab automated on a small scale.

In this particular application, the soft-lithography process produces a quick, inexpensive, high-quality product. The PDMS polymer is flexible, optically transparent (for applications where we want to use a laser to analyze our sample), impermeable to water, non-toxic to cells, and inexpensive. On the other hand, it *is* permeable to gases — which means we can't use it in any applications involving gas analysis. Also, its surface is hydrophobic (resists water), which makes it both prone to trapping air bubbles as well as susceptible to binding to proteins and cells — we can't do much analysis when our cells are always sticking to the wall! But, as described above, we oxidize the PDMS surface to bind it to another layer of PDMS or silicon. Fortunately, this oxidizing also renders the surface hydrophilic (attracts water), solving our problems of air bubbles and cell binding.

Since PDMS adheres well to itself, it can also be layered in more than just two layers, allowing for the creation of complicated fluid interconnects. Additionally, metal needles self-seal against PDMS, allowing syringes to be used to supply fluids or act as pressure reservoirs. George Whitesides and his team at Harvard have been working with PDMS and microfluidics for years, producing some rather ingenious innovations, including the soft lithography technique described above as well as pneumatic valves and micro mixers.

Moving honey

When you work at the micro-scale level, the particles in the fluid you're working with are going to be almost the same size as the device itself — a fact that changes the basic physics of fluids dramatically. Scientists obviously use some of these very same properties to move and mix fluids. Since our surface-to-volume ratio at the micro-scale level is much greater than at the macro scale, properties such as viscosity, surface tension, and electrical charges all play a great role. In the following sections we'll discuss these three properties in some detail — and, as an added bonus, we'll take a look at a unique way to move within this nano-realm.

Viscosity

Viscosity is the measure of resistance of a fluid — its "thickness," if you will. Fluids at the micro and nano level are highly viscous — they move more like honey than water. As a result, a property known as *laminar flow* comes into play, preventing fluids from mixing readily. With laminar flow, when two liquids come together they stay parallel to each other as they go down the channel, flowing smoothly and without turbulence.

Think of laminar flow and turbulence as polar opposites. If a real-life example helps, think of what happens when water flows from your sink faucet. First you have a steady stream of water that looks smooth, solid, and transparent, but once the stream hits the sink, the water breaks up violently, turbulence is created, and the water ceases to be transparent.

One nice thing about fluids at the nano level is that we rarely have to worry about turbulence — which is a big help when you have to do some number crunching. Laminar flow allows velocity calculations to be solved analytically instead of treating it as the average behavior as we do when working at the macro-scale level — the level where turbulence is once again involved.

Agilent Technologies and Caliper Technologies have collaborated to exploit laminar flow to help speed drug discovery — technology that they're using today. Both came up with techniques where you could first mix cells together with a fluorescing dye and then observe how the mixture interacts with the different chemical compounds you've identified as the stuff that could potentially become new drug candidates. The dye glows when the new drug reacts with proteins on the cell's surface — a sure sign that the drug can enter the cell so that it can (potentially) cure the disease being targeted.

The trick here is getting your cells to line up nicely in the microchannels of your lab-on-a-chip — think well-behaved children in the school lunch line here — so that each individual cell could be observed separately. Now, it turns out you can't simply manufacture microchannels so small that only one cell at a time makes its way through — if we tried to created a small nano-size channel just big enough for a single cell to go through, all the cells would hit it at once and quickly plug it up. But, if we introduced a buffer fluid that pinched our cells against the wall (thank goodness for the laminar flow principle), we could effectively have all the cells line up in single-file formation, as shown in Figure 10-3. Such a buffer fluid forces the cells to pass through the channel one by one, where a laser beam could do its magic. Those cells that fluoresce and are detected by the laser indicate that the drug is effective.

Figure 10-3:
As cells move through the microchannel (left), a buffer fluid pinches them against the wall (center) allowing them to be read by a laser (right).

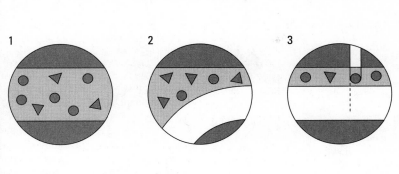

Now, if you've ever tried to stir honey, you know it's hard work. And if you've ever tried to mix together two viscous fluids such as honey, you know it's very hard work. Now try doing it with tweezers . . . okay, not really with tweezers but . . . well, sort of — keep reading. We've used laminar flow to pinch off and direct fluid — which is a good thing. Laminar flow, however, is also making it hard for us to mix stuff we actually want to mix — which is a bad thing. Mixing at the nano scale is governed by diffusion — over a long enough distance, the two fluids will diffuse into each other. As shown in Figure 10-4, however, it can take some time for diffusion to do its work. Here the two different colored dyes have run some distance side by side, and still haven't fully mixed.

Figure 10-4:
Over a long distance, two dyes will diffuse but may not entirely mix.

Red dye

Black dye

500 μm

80 μm

Fluorescence

If you've ever gone to a haunted house for Halloween, you probably noticed a lot of glowing objects around you that may have looked just a tad ghostly. Maybe even that white shirt you had on had an eerie purple glow. Now, the glowing objects were surely made of ectoplasmic goo, like that stuff in *Ghostbusters*, but the eerie white shirt was a result of the black lights on the track lighting hitting phosphors on your shirt — remnants of your tried-and-true laundry detergent — causing the shirt to fluoresce. (Okay, okay, we were kidding about the ectoplasmic goo. They were really extraterrestrial objects.)

Fluorescence is a property of some molecules — like phosphors, for example — that are able to absorb one wavelength of light and then emit light at a higher wavelength. The energy associated with our friend the photon determines how high up the wavelength spectrum we're talking about here, and, hence, the color — if in the visible spectrum. This energy from photons is transferred to electrons, causing them to become "excited" — not just "happy to see you," but also discharging excess energy just like when kids eat too much candy at Halloween.

Electrons in an excited state are unstable and tend to lose this excess energy. It turns out that, in fluorescent molecules, electrons lose this energy in two stages — a small amount at first (given off as heat) followed by a large amount which results in an emission of a photon of light. The energy of this emitted photon is lower than the absorbed photon, resulting in a difference in the color of light that is emitted.

The figure below shows a blue light with high energy exciting an electron sending it up to its "excited state." This electron is unstable and gives off small amounts of energy in the form of heat. Eventually, this electron gives up and goes completely back down to the ground state, its stable state, emitting a photon in the process. This emitted photon is lower in energy but higher in wavelength, giving off a green color.

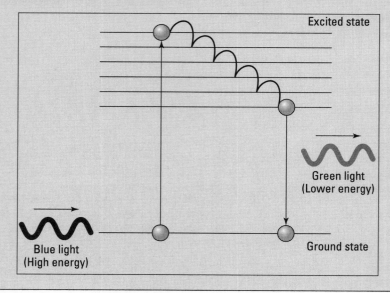

Excited state

Green light
(Lower energy)

Ground state

Blue light
(High energy)

Going back to our example of white shirts glowing eerily in Halloween haunted houses, black light bulbs generate ultraviolet light around 350nm — a short wavelength with high energy. In turn, the phosphors generate visible light between 400nm and 700nm — longer wavelengths with lower energy. We can't see the ultraviolet light emitted from the black lights but we can see the visible light from the fluorescence of the phosphors.

Fluorescence has three major advantages when it comes to medical applications: high sensitivity, high speed, and safety. The fluorescence signal is proportional to the concentration of the substance being investigated, giving it high sensitivity. A change in concentration can be detected in picoseconds, giving it high speed. Fluorescence is also noninvasive — safely monitoring samples, not harming living tissue, and not producing hazardous byproducts.

One way to get your two streams to mix is to narrow the streams as much as possible. That way, we only have to diffuse over the width of one stream. We do this by *hydrodynamic focusing* — we use laminar flow to pinch and create as narrow a stream as possible, as shown on the left in Figure 10-5. Another way to mix is to introduce turbulence — not especially easy, to be honest, but it can be done at this level. By adding irregular ridges on the bottom of our flow channel, as shown on the right in Figure 10-5, we can cause chaotic motion to occur in the cross section, leading to rapid mixing.

Surface tension

Okay, now that we have viscosity under our belts, we're stuck with just two more properties of fluids to work through. Number Two on our list is surface tension.

You might say to yourself, "surface tension, schmurface tension, what's the big deal?" But at the nano level, fluids and how they interact with various surfaces can become crucial. Imagine, for a moment, a water droplet on a waxed car. This water droplet has a bulk inside with a surface shell outside — kind of like an itty-bitty water balloon. This bulk inside has water molecules interacting just with each other, whereas the surface molecules interact with both each other and with air. The surface molecules have to expend more energy because they don't experience the same stabilizing hydrogen-bond interactions as the water molecules in the bulk. The larger the surface area, the greater the energy cost. This excess energy required to expand the surface by a unit area is what we call *surface tension*. Surface tension is the pull of a liquid into its most compact form to minimize the amount of energy used, keeping the surface area to a minimum — which is, come to think of it, precisely that force that pulls water into a droplet on a waxed or oily surface. Since we're dealing with small amounts of fluids, the surface forces overwhelm the bulk forces — and the little droplets keep their cute chubby shapes.

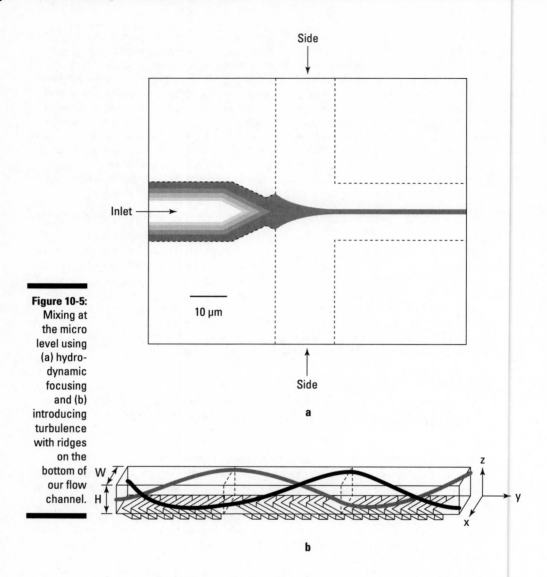

Figure 10-5:
Mixing at
the micro
level using
(a) hydro-
dynamic
focusing
and (b)
introducing
turbulence
with ridges
on the
bottom of
our flow
channel.

So, how can we get such surface forces to work for us? Funny you should ask. There are, in fact, many ways to control fluids by using surface interaction, including varying the temperature of the system and the electrochemistry of the surface molecules. One interesting way to control the surface area is with light — we can actually move a droplet with different wavelengths of a laser light. Scientists at the Tokyo Institute of Technology, for example, have used a UV laser to change the chemical structure of a droplet's surface — allowing them to change its surface energy and spread out the droplet; then they used a blue laser to bring the droplet tighter together. By varying the intensity at one end of the droplet, we can get the droplet to move, as shown in Figure 10-6. Here, the droplet is spread out with UV light while a blue light is introduced gradually — until (when the blue is at maximum) the droplet is restored to its original shape — *at a different location*. This process can also move a droplet within a tube, not just on a flat surface.

An interesting spin-off of using light to affect fluids involves *optical tweezers*. With optical tweezers, we can use a strongly focused laser beam to grasp and move small, translucent particles. Without going into too much detail, when a particle we want to grab comes in contact with our laser, there's a "restoring force" that pulls the object back to the center of the laser's beam. A lot of this has to do with the material of the particle and the type of laser wavelength used . . . not to mention, a lot of explanation of optics. The nice thing about using a laser is that we don't destroy or change our particle — great when dealing with something that's alive. We'll stop there and hope you're convinced that this works.

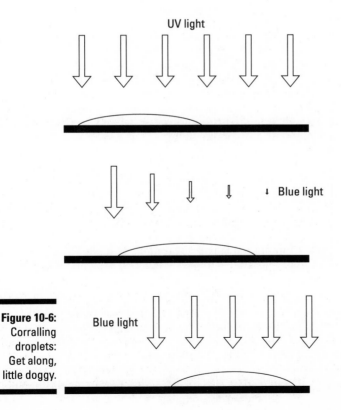

Figure 10-6:
Corralling droplets:
Get along, little doggy.

David Grier and his group at the University of Chicago have taken optical tweezers a step further with his "holographic optical tweezers." In Grier's work, a laser is split into sub-beams that can independently manipulate many tiny particles. Not only can these particles be moved and mixed but also set spinning. Individual cells can even be picked up and placed in a tube. Additionally, by changing the laser wavelength, the folks at Chicago can sort objects ranging from 10 nanometers (about virus size) to 100 microns. The fun doesn't stop there — holographic optical tweezers are incredibly flexible, able to sort cells by size, surface charge, magnetism, and shape — just by changing the laser's power, wavelength, or orientation.

Arryx, Inc., holds the exclusive license to holographic optical tweezers — a commercial product based on Grier's innovations. Using their first product, BioRyx 2000, researchers can manipulate hundreds of microscopic objects independently and simultaneously, in all three dimensions. The next Arryx product is set to be MatRyx, a high-throughput cell sorter used by the cattle industry to sort male and female reproductive cells for breeding — now *that's* a unique application.

Electrical charges

Two properties of fluids down and one to go. Get ready for a charge.

Another method for moving particles through our microchannels is electrokinetics — moving substances by applying an electrical charge (specifically, a voltage difference) across the microchannels, as shown in Figure 10-7. Okay, trying to move fluids at the macro scale by applying a voltage and expecting an impact would be ridiculous, but when you get to sizes smaller than 10 microns, it works — on particles.

Intrigued yet? There's more. Keep in mind that we have the following *two* types of electrokinetics:

- **Electrophoresis** consists of moving particles through our nanochannel by placing an external voltage difference across the entire channel — same as shown in Figure 10-7. This voltage difference moves charged particles through the channel. It turns out that we can separate these particles based upon charge and size — the big particles move slower than the small particles. This process is used today for DNA fingerprinting and general fluid composition analysis.

- **Electro-osmosis** is similar to electrophoresis in respect to using a voltage difference to move our fluid through a nanochannel. However, we also charge the sides of the channel wall. Charged particles in the liquid interact and slip past the charges on the channel walls. This allows us to push the fluid column down the channel with all the particles at the same speed. This looks like a slippery, square plug being pushed down the channel and is actually called "plug flow."

Figure 10-7:
Electrokinetics involves using a voltage difference across the channels to move particles by electric charge.

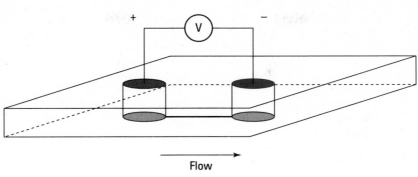

Flow

Thinking differently

Okay, we've gotten all of our fluid properties out of the way, but we can't put away the thinking caps quite yet. If we really want to make the whole microfluidics/lab-on-a-chip/microchannels thing work, we have to find a way to keep our samples from going down the wrong way (you know, what happens when you try to drink that can of soda while your friend tells you a great joke). PDMS, that polymer of choice for microfluidics, can be used to create a mechanical valve designed to accomplish just that task.

Figure 10-8 gives you a bird's-eye view of one way this can be done. We sandwich a thin layer of our PDMS between two other layers of PDMS, using a process called *multilayer soft lithography (MSL)*. (For more on soft lithography, see the "Fabrication through soft lithography" section, earlier in the chapter.) To glue the layers together, we oxidize the PDMS layers and press them together — but we don't completely expose the middle layer to the oxidation process. That means the "unglued" layer is free to move, allowing fluids to go through it but not come back through.

MSL was developed by Stanford biophysicist Stephen Quake — the guy who co-founded Fluidigm Corp. This layering process allows Fluidigm the flexibility necessary to design not only valves but also unique pumps and complex microchannels — giving them a competitive edge in this emerging market.

Keeping things moving

Now that we have some idea as to how we move honey, how do we move *in* honey? Since we have no turbulence — the stuff that allows for gliding — the conventional notion of swimming is thrown out the window. Without turbulence, making a single motion and exactly reversing it will put you right back where you started. Therefore, the trick is to design something that is not identical in the reverse direction — something cyclical like the flexible oar or corkscrew you see in Figure 10-9. (These look a lot like flagella, the whiplike organelles that unicellular organisms such as bacteria use to propel themselves about.)

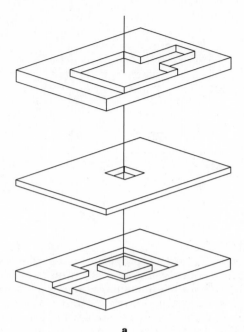

a

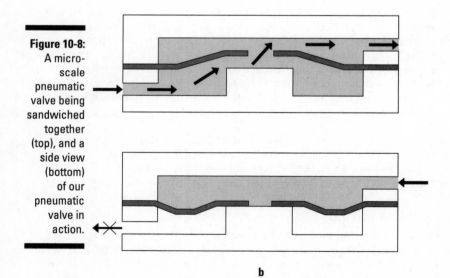

Figure 10-8:
A micro-
scale
pneumatic
valve being
sandwiched
together
(top), and a
side view
(bottom)
of our
pneumatic
valve in
action.

b

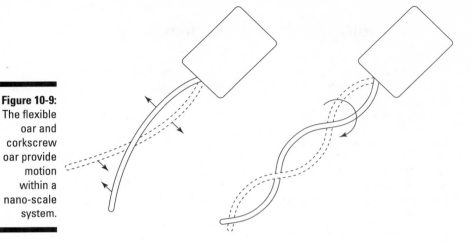

Figure 10-9:
The flexible oar and corkscrew oar provide motion within a nano-scale system.

These biomotors are made of complex "motor proteins" which use ATP (adenosine triphosphate) as an energy source. Carlo Montemagno and his group at Cornell University have incorporated molecular motors into engineered devices. They've used a biomolecular motor that runs on ATP, and attached a base to the bottom and a nickel nanowire to the top to act as a propeller, as shown in Figure 10-10. Amazingly enough, these work as advertised.

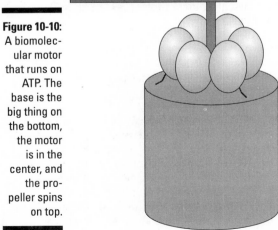

Figure 10-10:
A biomolecular motor that runs on ATP. The base is the big thing on the bottom, the motor is in the center, and the propeller spins on top.

Giving cells the squeeze test

Getting lab specimens to line up nicely and move along (no talking, please!) through the microchannels of our lab-on-a-chip is a great accomplishment, but it's only half the battle. We still need to come up with an effective way to give specimens the once over — to analyze them, in other words. In our discussion of laminar flow (back in the "Moving honey" section earlier in the chapter), we highlight one way of carrying out such an analysis — using a laser to count cells that have fluorescent tracers attached to them. Another approach, which likewise uses a laser to analyze a cell, is called an *optical stretcher*. After the cells are lined up, one-by-one, they are squeezed individually with a laser to test their elasticity. It turns out that cancer cells are softer than healthy ones — and this optical stretcher can detect a clear difference between the two more quickly than current elasticizing methods. Optical stretching can sample several hundred cells per hour, as compared to ten cells a day — a huge advantage — thanks to nanotechnolgy.

Biosensing with nanowires

So, now we have our lab-on-a-chip and know how to move fluids through them. Previously, we described using fluorescence to illuminate specific cells to be read by our laser. (See the appropriately named "Fluorescence" sidebar, earlier in the chapter.) We think we can do better than this and cut out the fluorescence step. Imagine if you will that we can incorporate nanosensors directly into the lab-on-a-chip — as the fluid went over the sensors, the sensor detected the particles, replacing the need for the fluorescent dye and laser. This sensor will have to react to biological molecules (proteins) — we'll call this a *biosensor* (not very original).

This sensor will also have to be small — from our tool chest of materials described in Chapter 3, we have nanowires (they're small). So, let's make a biosensor out of nanowires and incorporate them into our lab-on-a-chip (sounds like fun).

Nanowires, because of their small size, make very sensitive devices. We're going to use a nanowire as a component of our tiny biosensor. This nanowire (any nanowire will do) will be turned into a transistor, in much the same way as carbon nanotubes were in Chapter 6: The researchers bring each end of the nanowire into contact with other metal wires so they may pass current through the nanowire. Now, if the nanowire was a traditional transistor, an electrode would be placed close to it, altering its conductance so it could be turned on or off. In the case of our biosensor "transistor," however, the molecules that we're trying to detect are what change the conductance — they actually cause the transistor to switch, so there's no need for a switching electrode. Keep in mind that our biosensor "transistor" isn't a transistor in the computer sense, but functions in a similar way.

How do we pull this little trick off? We coat the nanowire with antibodies that stick only to the proteins that we're trying to measure — not to other molecules. When these proteins bind to the antibodies, they interact with the electrons traveling in the nanowire's surface layer — which alters its conductivity. If the nanowire is small enough, the surface layer is pretty much the entire wire — and with very little inside bulk, a large portion of the atoms making up the wire are on the surface. This allows the nanowire to be very sensitive; a minute amount of change to the surface would significantly change the conductivity. If the nanowire is small enough, it may even be able to detect individual molecules.

Charles Lieber's group at Harvard demonstrated that nanosensors like these can detect a single virus particle. Given this sensitivity, we could conceivably use such biosensors to detect a wide range of diseases by their early warning signs — and treating the disease sooner means a greater chance of survival.

At the top in Figure 10-11, you can see a piece of nanowire between the metal leads, with Y-shaped antibodies attached to the surface. The molecules, proteins, or other biological species bind to the antibodies, which changes the conductance of the wire. These nanowires can be placed in arrays — with each wire coated with a different antibody (representing a different detector), as shown at the bottom in Figure 10-11.

Carbon nanotubes: Mr. Sensitive?

Carbon nanotubes have established themselves as chemical sensors (as we discuss in Chapter 8), but they are also a hot prospect for use as biosensors. Imagine replacing the nanowire we've been describing here with carbon nanotubes between the two metal contacts: Antibodies are attached to the nanotubes and the conductance is measured — a quick way to detect specific molecules in the environment. And as if that weren't enough, a couple of other architectures for biosensors take advantage of carbon nanotubes:

✔ An individual nanotube can be attached to the end of a microtip providing nanometer

resolution able to probe single molecules of DNA or proteins. This allows scientists to take electrical, electrochemical, and electrophysiology measurements of individual cells.

✔ Nanotubes can be randomly distributed so they form a flat sheet. This thin film is then placed between two electrodes, forming our biosensor. This approach circumvents the high cost of making our single nanotube biosensor — those things are small and we haven't found anyone with the patience to lay them out one at a time . . . yet.

James Heath has been working on this technology and combining it with the lab-on-a-chip idea. Dr. Heath has been studying nanowires for use as biosensors, collaborating with Stanford's Stephen Quake, a co-founder of Fluidigm Corp. (Quake's work on creating lab-on-a-chips using multilayer soft lithography got an honorable mention from us earlier in the chapter.) In this case, fluids are shuttled down microchannels in single file, and positioned over these nanosensor arrays where they can be studied one at a time — another way to analyze our sample.

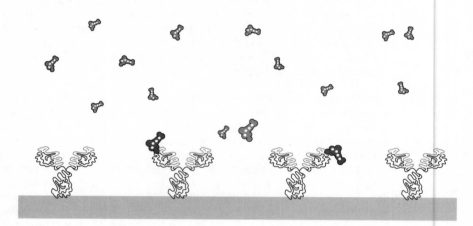

a

Figure 10-11:
Top: a single-nanowire biosensor. Bottom: an array of nanowires detecting multiple molecules at once.

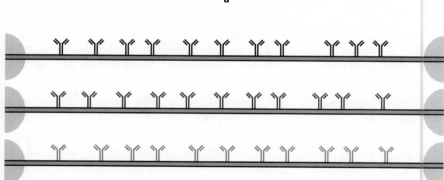

b

Super X-Ray Vision

Any lab-on-a-chip we could come up with takes small biological samples and analyzes them quickly so that what ails us can be diagnosed in a heartbeat, allowing for faster treatment. As it stands, all this lab-on-a-chip analysis takes place outside the body — a good start, but not the radical medical break-through that's going to get us all on the cover of *People* magazine. What we *really* want is to be able to tell whether our drugs are working within the body — and that means imaging biological processes nondestructively in the body, in real time. Such a medical application would allow us to monitor dis-ease processes and measure how well the therapy is working. We'd even be able to study how diseases disrupt normal molecular and cellular signals and pathways and how individual molecules work inside cells.

This *molecular imaging* — the official term for snooping in on the private lives of molecules — has taken root at Massachusetts General Hospital's Center for Molecular Imaging Research. Using fluorescent tags, radiologist Umar Mahmood traces the paths of destructive enzymes secreted by a tumor. His computer screen helps him track, and eventually determine, the underlying causes of this disease. Mahmood has worked with chemists to develop "smart probes," allowing the tag to give an indication when it meets its target. In this case, untreated tumors fluoresce brighter than treated tumors in real time. This may not seem like a big deal, but current methods involve waiting a few months to see whether the tumor physically shrinks, indicating suc-cess. Studies along these lines will eventually lead to detecting cancer indica-tors — molecular aberrations — that precede physical symptoms by months or even years. This procedure could even replace the biopsy (in which a sur-geon cuts out a piece of tissue to subject to tests — not pleasant).

Molecular imaging comes in a variety of forms. One involves *dendrimers,* polymer-based treelike structures that are as little as 3nm across. Researchers at the University of Michigan Center for Biologic Nanotechnology are using dendrimers to deposit a type of DNA into the cell making it glow. At Ben-Gurion University in Israel, Raz Jelinek affixes color-changing polymers to human cells. These polymer coatings are arranged in 30–150nm patches, which fluo-resce if something perturbs the cells' membranes. This little fluorescing "alarm bell" lets drug developers determine if drugs are reaching their targets and lets medical researchers see if viruses are mounting attacks.

Tracers in fullerenes

Buckyballs (fullerenes or C_{60}, described in Chapter 3) may be used as contrast agents for medical imaging. Contrast agents improve the resolution of the image by increasing the brightness where the agent resides. Fullerenes — those cute soccer-ball-shaped molecules — act as "cages" for contrast agents, lowering any residual toxic side effects. This results in a *metallofullerene* — a metal consisting of only a few atoms encaged within a fullerene (a metallofullerene can be seen on the cover of this book). It's important to note that fullerenes are small enough to penetrate the *blood-brain barrier* — the membrane that protects the central nervous system from toxins and infection that may be found in the rest of the body.

In magnetic-resonance imaging (MRI) — the current high-tech benchmark in medical imaging — gadolinium (a metal) serves as a contrast agent. Gadolinium has very few harmful side effects and low metal toxicity, but the industry is always looking for ways to prevent metals from leaving the delivery vehicle and causing further harm to the body. Research involving gadolinium housed in a fullerene (C_{82}) suggests that the metallofullerene provides a better contrast agent than gadolinium alone. It's a win-win-win result: better contrast, lower toxicity, and faster scans.

The tiny fullerene cages also have a use in nuclear medicine. They offer an alternative to using radioactive metals alone as contrast agents (not ideal because they're absorbed by the body, where they aggregate in tissues and remain there for the life of the patient). Fullerene cages can be used to trap these radioactive metals — the radioactive metal stays in its cage-molecule, where it can be normally filtered and expunged from the body.

Modifying the fullerene cage itself may provide some benefits. Lon Wilson's group at Rice University (for example) has modified the gadolinium-based metallofullerene (C_{60}) so the metallofullerenes aggregate — which enhances their image properties. Companies like C Sixty modify the cage to make the molecule soluble and accessible to cells, providing further uses (such as an antioxidant which inhibits oxidation and helps limit cellular damage caused by free radicals, highly reactive molecules).

Quantum dots

We describe quantum dots, semiconducting nanocrystals roughly 5nm in size, in Chapter 8, where we show how they fit into the electrical world. They also have applications in the biological world as fluorescent tags. Quantum dots are nanometer-scale nanocrystals composed of a few hundred to a few thousand semiconductor atoms made out of bio-inert materials — meaning they are nonintrusive and nontoxic to the body. Additionally, unlike fluorescent dyes (which tend to decompose and lose their ability to fluoresce), quantum dots maintain their integrity withstanding more cycles of excitation and light emission before they start to fade. Changing their size or composition allows scientists to cater their optical properties — which means they can fluoresce in a multitude of colors. This effect is called *quantum confinement* (hence the name *quantum dots*) — they have quantized, discrete energy levels that are directly related to their size.

Interestingly enough, quantum dots can even be tuned to fluoresce in different colors with the same wavelength of light. In other words, we can choose quantum dot sizes where the frequency of light required to make one group of dots fluoresce is an even multiple of the frequency required to make another group of dots fluoresce; both dots then fluoresce with the same wavelength of light. This allows for multiple tags to be tracked while using a single light source.

A. Paul Alivisatos and his company (Quantum Dot Corporation) have used these concepts in their Qdot product — a quantum dot surrounded by an inorganic shell that amplifies its optical properties while protecting the dot from its environment. The Qdot can have a variety of attachments to its shell, allowing it to attach to specific cell walls — or even penetrate a cell and light it up from the inside. In the summer of 2003, Quantum Dot Corporation joined forces with Matsushita Electronic Industrial Co. (Panasonic) and Sumitomo Corporation Biosciences to develop advanced optical and image-processing technologies that utilize the Qdot. Products under this agreement are expected to generate revenue of more than $100 million per year for Quantum Dot Corporation by 2007. (Tiny product, big bucks.)

An example of quantum dots in action involves targeting and imaging cancer cells. Researchers at Emory University, Georgia Tech, and Cambridge Research and Instrumentation have used quantum dots to identify tumors in mice. These quantum dots were made of cadmium selenide-zinc sulphide, each 5nm in diameter. They were coated with polymers to prevent both the body from attacking the quantum dot and to keep the dots themselves from leaking toxic cadmium and selenium ions. (Eek! Toxic! Read on. . . .) Finally, they attached antibodies to the outer shell that was first targeted and then attached themselves to a prostate tumor cell surface. The scientists injected the quantum dots into the circulatory system and the dots accumulated at the tumor, which could then be detected by fluorescence imaging. As an added benefit, these quantum dots have a large surface area allowing for a dual role of both diagnostics and therapy — the surface is big enough to attach both diagnostic and therapeutic antibodies to the surface.

I know what you're thinking — "Rich, what's with all this about toxicity and *eek*-ing? We can't have our imaging material causing more problems." And, you're right. The scientists have also thought of this and have done experiments mirroring the body's environment to make sure quantum dots and their coatings are stable over a broad range of pH and salt conditions — even hydrochloric acid. And they passed with (ahem) flying colors.

Carbon nanotubes also have this fluorescence quality. R. Bruce Weisman and his group at Rice University have determined that semiconducting carbon nanotubes fluoresce in the near-infrared spectrum and can be fine-tuned to different wavelengths by varying the nanotube diameters. The near-infrared spectrum is particularly important for biomedical applications — but nothing in the human body fluoresces in this region of the spectrum; in effect, human tissue is fairly transparent. It's especially handy that nanotubes also maintain their fluorescent properties inside living cells, with no adverse effects to the cell. (For more on carbon nanotubes, see Chapter 4.)

Down the road, such carbon nanotube technology may be used along the same lines as the quantum dots — you could end up wrapping the tubes with a specific protein, allowing them to target cells (such as tumors). Along these lines, a proposal by Michael Strano and his team at the University of Illinois-Urbana/Champaign (involving a glucose-detection optical sensor) looks especially promising. Here, nanotubes are wrapped in a glucose oxidase and placed inside a small, porous capillary (200 microns across by 1cm in length). The capillary pores are only big enough for glucose to penetrate; once through, the glucose promptly reacts with the oxidase solution — changing the fluorescence properties of the nanotubes. This capillary is subsequently inserted just underneath the skin, but within range of being able to detect the near-infrared flourescence. Imagine a patient with diabetes wearing a watch that periodically checks the fluorescence/glucose and sounds an alert if levels are too low or too high — all without needles.

Unlike quantum dots, nanotubes don't contain heavy metals, so they don't raise any toxicity issues. Additionally, nanotubes can be fine-tuned to very narrow wavelengths, providing fluorescence in a greater number of wavelengths, giving us greater flexibility (in other words, more colors to our palette). Such properties may give nanotubes the advantage among products marketed as laboratory imaging markers — we'll have to see whether they can squeak by to take the lead from quantum dots.

Mapping Our Genes

Deoxyribonucleic acid (DNA) is the nucleic acid carrying the genetic blueprint of all forms of cellular life. You may remember from high-school biology class that DNA has four building blocks (called *bases*) abbreviated A, T, G, and C. Each of these bases only pairs up with its complement — A only matches with T and G only matches with C — to form the famed double helix, as shown in Figure 10-12.

One nice thing about DNA involves a property known as *hybridization* — DNA will bind only to a complementary strand, and to nothing else. "And why is that handy?" you ask. Well, imagine you want to detect a structure that contains the following strand: ACTGTA. You can do so by creating the complementary strand (TGACAT) and, when they hybridize — creating a double-strand — they'll be bigger and you can measure this with sophisticated instruments.

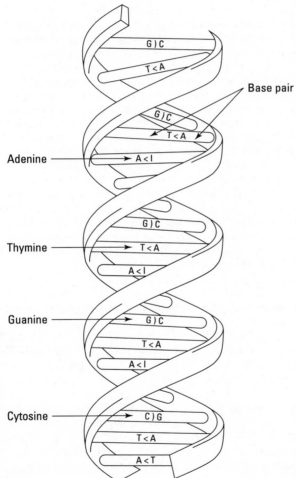

Figure 10-12:
The DNA double helix, with bases Adenine (A), Thymine (T), Guanine (G), and Cytosine (C).

Don't get me wrong — this isn't easy and is usually done with many double-strands being formed and detected. (Nanotech will help bring down this sample size — read on.) For our example of six bases — made up of the four different possible bases — we have 4^6 (4,096) possible combinations. Then comes the voice from the back of the room: "Hey, Rich! Why is that important?" Well, if we know the DNA sequence of a unique short section (say 15 bases) of a disease, we can sense this section of the strand by using the complementary strand. What we get is a one-in-one-billion (4^{15}) chance of error per strand tested — very accurate, in other words.

It's nice to know that DNA carries the genetic blueprint of all life, but what can you actually *do* with it? It's not as if you can use such a blueprint to build a summer home, right? Actually, there are a few reasons for wanting to know the genetic blueprint of an individual. For openers, you can determine a subject's likelihood for developing certain diseases — and determine which drugs will be most effective at treating those diseases. This second reason is called *pharmacogenetics,* which is projected to be a $30 billion industry by 2013, treating cancer, immune disorders, HIV/AIDS, and neuropsychiatric disorders. Neuropsychiatric disorders are set to benefit greatly, particularly because they are difficult to diagnose and treat.

Given the stakes — and money — involved, it's clear that we need to find easy, foolproof ways to map our DNA. Current methods use fluorescent dye markers, but nano-approaches point us in new directions. One new way to map our genes involves using a lab-on-a-chip, where some of those strange fluid properties of microfluidics (which we discuss earlier in this chapter) could be used to separate proteins and DNA. Additionally, a drop of blood can be instantly checked against known viral or bacterial DNA and water supplies could be checked for toxic microorganisms.

Nanogen and Orchid Biosciences are two lab-on-a-chip companies exploring DNA detection.

Another way to map our DNA is to utilize some of the same optical concepts we discuss in the "Quantum dots" section, earlier in this chapter. DNA is attached to gold or silver nanodots (13nm wide) that are suspended in a liquid. Each gold particle has the same base pair — but when a linker (such as Anthrax DNA) is introduced, the gold particles form larger clusters, which changes their optical properties, as shown in Figure 10-13. (Here particle size dictates the wavelength of light, hence color.) Imagine that without the linker, the liquid looks purple; with the linker, however, the liquid looks red — providing a quick macroscopic analysis. Because of this color change, these are called *colorimetric sensors.*

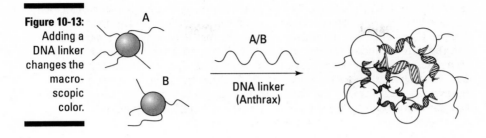

Figure 10-13:
Adding a
DNA linker
changes the
macro-
scopic
color.

A

B

A/B

DNA linker
(Anthrax)

Microarray

The DNA microarray is a veritable lawn of single-strand DNA molecules that stand on a postage-stamp-size chip. Microarrays are used for many different purposes:

- ✔ to identify DNA sequences and subtle differences in gene composition
- ✔ to determine gene response to potential new drugs
- ✔ to study evolutionary history of genes to determine growth processes

If all that sounds a bit off the beaten path, guess again. The DNA microarray market — including protein and cellular microarrays — is slated to reach $1.6 billion in 2008. The microarray chip depicted in Figure 10-14 shows a checker-board array, where each section contains thousands (even millions) of copies of the same DNA strand.

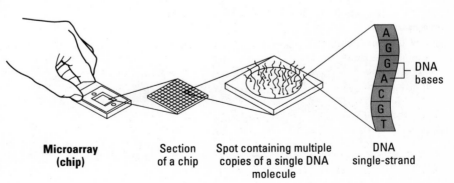

Figure 10-14:
A DNA
microarray
chip is
composed
of a
checker-
board array
of millions
of single-
strand DNA.

**Microarray
(chip)**

Section
of a chip

Spot containing multiple
copies of a single DNA
molecule

A
G
G
A
C
G
T

DNA
bases

DNA
single-strand

When a sample is dropped onto the microarray, as shown on the left in Figure 10-15, every section of the array performs the reaction at the same time — in parallel. The sample contains fluorescent tags that indicate whether the strands match the specific, single strand of DNA on which the array is based. The microarray chip is then read into a computer, which quantifies gene activity on a color-coded readout, as shown on the right in Figure 10-15. Further computer analysis helps determine whether the sample is a good drug candidate or not.

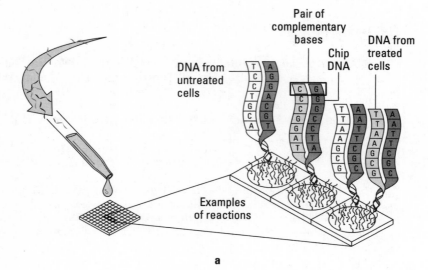

a

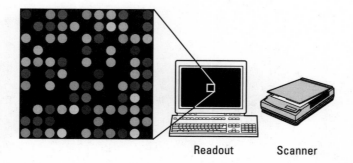

Readout   Scanner

b

A major use of these microarrays (besides customizing medicine to fit patient needs and determining how vulnerable a patient is to disease) is in building a detailed picture of the effect of a drug or disease, a technique called *expression profiling*. It's not enough to just know if a genetic blueprint is present; it's equally important to determine the amount of proteins that are produced by that blueprint. By knowing various levels of proteins, scientists can gain detailed "snapshots" of how a cell's functions have been altered by a drug or disease. This snapshot reflects the molecular state of the sample under specific conditions — and that knowledge can help determine treatment.

All of this investigation and intervention at the nano level will help us in our ultimate goal: determining how genes actually work. We know that genes work together or in series to induce a cellular response. We also know that some of these genes are associated with each other — knowing precisely how one gene operates sheds light on how associated genes may act in a similar situation. Additionally, knowing the associated gene may lead us to discover proteins not previously known to play a role in diseases.

Working on the DNA chain gang

In 2003, Jene Golovchenko and his team at Harvard developed a unique method for eventually reading DNA sequences. First they use an ion beam to poke a 3nm hole into a 10nm-thick silicon nitride membrane. Then they thread the double-stranded DNA through the nanopore, as shown in Figure 10-16, and measure the ionic current. The presence or absence of *current blockades* — that is, dramatic drops in the current that suggest an obstacle — describe individual molecular features like the sequencing of our DNA.

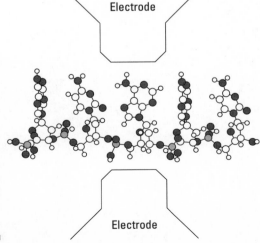

Electrode

Figure 10-16:
A single-strand DNA molecule, going single file between two electrodes embedded in a nanopore.

Electrode

Actually, studying molecules by using electrical signals across nano-scale pores is not new. *Biopores* — membranes utilizing a lipid bilayer (similar to a cell wall) in which we can control and insert nanopores — have been around since at least 1992. Biopores, however, do have some limitations: The pore size is fixed — and if the pores are destroyed at extreme temperatures, voltages, or pH conditions, there go your results.

By using the more stable solid-state silicon nitride membrane instead of a lipid bilayer, Golovchenko observed new phenomena — such as *folding* (a folded molecule's bend went through the hole first) and *pairing* (two connected molecules went through the hole at once). Although these phenomena don't sound spectacular, they give scientists a way into playing with molecules outside the natural states they're found in. In other words, if we bend a molecule in a way it doesn't normally bend, we may observe new and unique properties for that molecule. The ultimate goal for us (and Golovchenko) is to rapidly sequence DNA, providing a better understanding of our genes, to prevent disease and usher in a medicine catering to individual needs.

Chapter 11

The Fantastic Voyage into Medical Applications

"Man is the center of the universe. We stand in the middle of infinity between outer and inner space, and there's no limit to either."

from Fantastic Voyage *(1966)*

In the 1966 film *Fantastic Voyage,* audiences were introduced to their own bodies from the inside. A group of doctors got into a submarine-like ship and were shrunk down to the size of a cell. They were then injected into a human body to search and destroy a blood clot. Audiences got a fairly realistic look at some of the inner workings of the human body. (If you don't know what we're talking about, you may remember Dennis Quaid, Martin Short, and Meg Ryan in the 1987 movie *Innerspace* — same concept.)

It's a recurrent theme in science fiction: Human beings want to live long, healthy, pain-free lives — and we will continue to create (and invest in) ways to do so. Nanotechnology is an emerging reality that can help us along that path. It won't enable humans to shrink, but it can, however, help us modify and create particles that circulate through the body with as much control as if we were in there calling the shots. This chapter explores some unique approaches to delivering drugs and killing cancer that nanotechnology makes possible. We start with some of the problems pharmaceutical companies have with drug development — and discuss a "nano-way" to deliver drugs that addresses this problem. We then provide a recipe for *cooking* cancer (that is, eliminating it) and tell you about some new concepts of artificial blood and bones.

Understanding How Pharmaceutical Companies Develop Drugs

Bioavailability refers to how well a treatment can target specific cells. For the last hundred years or so, pharmaceutical products have suffered from poor bioavailability — their main approach is to flood the body with the drugs that are needed. One aspirin cures a headache but one hundred aspirins kill the patient (permanently "curing" the headache). This is especially not good for cancer and chemotherapy; increasing the amount of toxic drugs eventually kills the patient.

As with any business, companies doing drug development have to choose between risk and reward. The amount of risk in drug development is incredibly high — in terms of both time and money. The starting point is to determine which molecules of what compounds will be most effective at curing a disease. Different models of how the disease works are used — and these days computer programs speed up the process — but taking a new drug from research to development to administration still takes a long time. Even today, the average time between the patent application and the marketing of a new medicine is 12 years. And a patent expires after only 20 years. To top it off, some research projects are scrapped as late as advanced-stage clinical trials (roughly nine years after kickoff!).

According to Pharmaceutical Research and Manufacturers of America, only five candidate medicines are sent to clinical trials for every 5,000 drugs tested in the preclinical phase (that is, before they're approved for clinical tests). And of those five, only one is finally approved for release to the marketplace. Figure 11-1 shows the timeline of drug development from inception to market to patent expiration.

With high risk comes high cost — the price tag for research and development (for example) more than doubled worldwide between 1990 and 2000. In 2001 alone, a total of $43.63 billion was spent — a billion dollars here, a billion dollars there — pretty soon, you're talking about real money. Every new drug costs an average of $802 million to bring to market (which includes the cost developing those that didn't make it to market). The high rate of failure, the cost of clinical trials, and the time required for product approval all contribute to the growing cost of research. Result: Even though we're spending more money, we're producing fewer new medicines.

Only 32 new drugs were introduced in 2000 — a pretty low figure — and the number of new drugs developed yearly is decreasing while the cost of developing a new drug is increasing. This is not a good tradeoff, but there may be some light at the end of this particular tunnel: Rapid advances in biotechnological applications. Biotechnology patent applications increased from 590 in 1996 to 1,615 in 2000, and this trend can help move us in the direction of faster, less-costly drug development. The sequencing of the human genome, for example, is one new body of knowledge that can help us determine which medicines work by singling out the particular genes that contribute to illness. Nanotechnology, in conjunction with biotechnology, will aid in more effective medicines while lowering the costs.

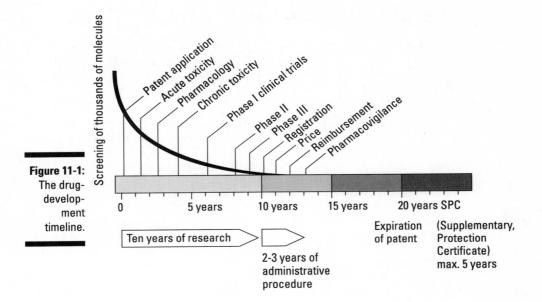

Figure 11-1: The drug-development timeline.

Delivering a New Drug the Nanotech Way

Our ultimate goal is to find a way to get drugs to a specific site in the body — that is, to increase bioavailability. Here are the steps involved:

1. Develop a delivery mechanism that packages the drug.

2. Insert the drug into the bloodstream.

3. Release the drug at the specific point of interest — for example, an infected body part or a cancer cell — for maximum effect.

Think of the delivery mechanism as the "submarine" in *Fantastic Voyage,* and the drug as the people inside aiming a "laser" to destroy (say) a blood clot.

Imagine the possibilities to replace current chemotherapy, an arduous process by which "anticancer" drugs are pumped through the body — killing not only cancer cells but also healthy cells — causing severe side effects. If we can deliver these drugs directly to the cancer cells, this will limit their effect on the healthy cells. This benefit is our motivation.

In this section, first we take a look at the building blocks of the delivery mechanism, explain how they congregate and form into the mechanism, and then discuss how they can be used to deliver the drug.

Oil and water don't mix

The building blocks to our delivery mechanism are molecules called *amphiphiles* — *amphi* from the Greek word for "both," and *phile* from the word for "loving." Such a molecule "loves" both oil and water — and it has two distinct parts:

✔ Head: This group of atoms is *hydrophilic* (water-loving).

✔ Tail: This group of atoms is *hydrophobic* (water-fearing/oil-loving).

Figure 11-2 illustrates how such a molecule is arranged.

Yeah, we know, "hydrophobia" used to be a synonym for "rabies" — but what's going on here is a chemical process, not an infection.

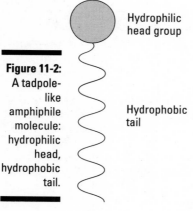

Hydrophilic head group

Hydrophobic tail

Figure 11-2:
A tadpole-like amphiphile molecule: hydrophilic head, hydrophobic tail.

Okay, so what are amphiphiles good for? One great application is laundry detergent: Since oil and water don't mix, oils from our skin that are deposited on our clothes would rather remain on the fabric than dissolve in the water. Soap molecules with hydrophobic tails that are soluble in oil therefore become embedded in those tiny grease globs. As you can imagine, given enough soap molecules, the hydrophilic heads surround the grease; the hydrophobic tails duck inside the grease — leaving the grease glob water-soluble. (These types of amphiphiles are called *surfactants* — they're "surface-active," performing their molecular duty at the interface of two different substances.) When the grease is surrounded by a water-soluble coating, the entire glob can dissolve — which removes the stain from your clothes.

Amphiphiles do dissolve in water but they prefer to keep their hydrophobic tails away from the water molecules. Not only do they congregate near the surface with their tails off the water and their heads down, but they also group together to create spherical structures called *micelles* — which leads us right into our next section.

Micelles (your cells?)

Say we continue to add amphiphiles — in this case, laundry detergent — into water in a washing machine. Since the detergent molecules' hydrophobic tails want to keep away from the water molecules, the detergent molecules congregate near the surface — with their heads in the water and their tails up in the air. Eventually we've saturated the surface of the water with detergent molecules; they look elsewhere to keep their hydrophobic tails from interacting with water. As you might guess, they tend to group together into spheres with their hydrophobic tails inside and their hydrophilic heads outside (nicely illustrated in Figure 11-3). This is the same concept when the detergent surrounds a grease glob, but this time, without the grease — the tails are what attract the detergent molecules together.

These spherical groupings of amphiphile molecules — called *micelles* — are created when enough amphiphiles are present that they reach a *critical micelle concentration* — so many that they feel compelled to form micelles instead of just congregating toward the surface of the liquid. The nice thing about these nano-size capsules is that they create themselves — put enough amphiphiles in one place, and you get them automatically. The nature of the amphiphiles (every one has one hydrophobic side and one hydrophilic side) is the driving force behind their organizing themselves into micelles. That's especially handy, given their small size — it would take a very long time to stick 'em together individually (I'm certainly not going to do it). The organization is very give-and-take — amphiphiles come and go within the micelle.

Micelles come in a variety of orientations. If we introduce amphiphiles into an oily liquid — as opposed to introducing them into water — micelles are still created, but now you get hydrophilic heads on the inside and hydrophobic tails on the outside — a reverse micelle, as shown in Figure 11-4.

At high concentrations of amphiphiles (higher than the critical micelle concentration), micelles may form *bilayers* — a reverse micelle surrounded by a regular micelle, as seen in Figure 11-5. These one-within-another micelles are called *vesicles* — which have an intriguing (if sketchy) resemblance to biological cells.

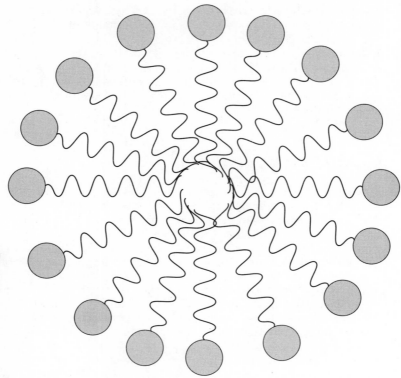

Figure 11-3:
A micelle
with the
hydrophilic
tails
grouped in
the center.

The soap molecules described earlier in this chapter represent one type of amphiphile. Human cell walls consist of natural amphiphiles called *phospholipids*. They look sort of like mutant soap amphiphiles — each molecule has *two* hydrophobic tails instead of one. Like micelles, the phospholipid amphiphiles are loosely bonded to each other and move about the shell of the vesicle. That's where the resemblance to biological cells gets sketchy: Not all cellular walls are as mobile — some animal cells are rigid, and red blood cells have a weblike protein skeleton — and such structures and properties are unique to living cells.

Micelles have another useful industrial characteristic: They can be any of various sizes. The internal cavities of small micelles are mostly made up of the hydrophilic tails (as in our original description of a micelle); larger micelles not only have their tails in the center but also include some water-insoluble substances in there (for example, that hapless grease glob). These sizes are *tunable* — by varying what's in the liquid, scientists can dictate the size of the micelles that form. They can all be the same size or they can be of various sizes.

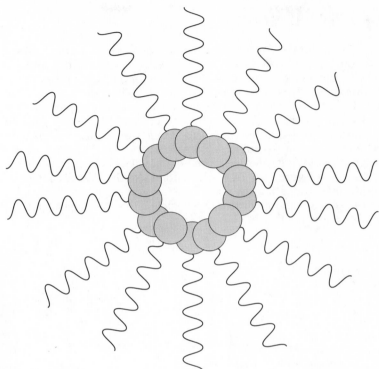

Figure 11-4:
A reverse micelle with the hydrophobic heads in the center.

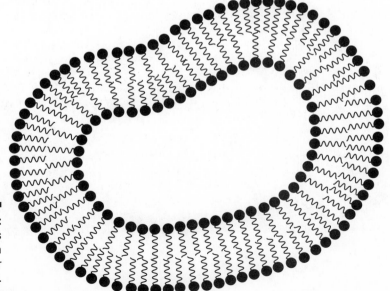

Figure 11-5:
Amphiphiles creating a bilayer (vesicle).

Special delivery

When amphiphiles form into a vesicle, you have a microscopic structure that has some handy uses — for example, as a package for delivering drugs to specific locations in the body. Recall that amphiphiles form only weak bonds among themselves — for the most part, they're free to move around within the shell of the vesicle. Individual vesicles can break apart into two smaller vesicles or join together into a bigger vesicle. That means you can inject a substance into — and extract a substance from — the vesicle. *Endocytosis* (*endo* — into; *cyto* — hollow vessel) is a process whereby cells absorb particles by enveloping them with the help of vesicles formed from the cell wall, as shown in Figure 11-6. (It works kind of like the ravenous glop in that Creature Feature standby, *The Blob.*) Conversely, *exocytosis* is the removal of particles by enveloping them in a vesicle and releasing them *outside* the cell wall, as shown in Figure 11-7.

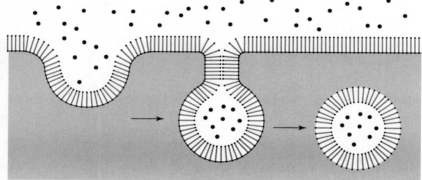

Figure 11-6: Endo-cytosis — molecules being encapsulated in a vesicle.

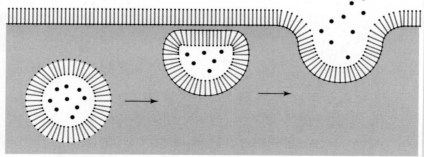

Figure 11-7: Exo-cytosis — molecules encapsulated and being released.

What we have here is a way to grab microscopic things and move them around. These handy bilayer vesicles act as *liposomes* — cells used to deliver drugs or genetic material into a cell. To deliver drugs, for example, you pre-pare the liposomes outside the body, encapsulating drug molecules in them. You then deliver the liposomes as micro capsules into the bloodstream, where they make their way to the target cells and release the drug. You can even determine what kind of cellular interaction goes on between the lipo-some and the cells by varying the type of phospholipids used.

Following are the three main methods that liposomes use to deliver drugs to a cell:

✔ Liposomes may stick to the surface of the cell walls, slowly diffusing the drugs into the cell. (See the part labeled "a" in Figure 11-8.) This method is good for consistent, time-released drugs — varying the structure of the liposome can vary its willingness to stick, vary the drug's release time, or both. A controlled release can respond to the needs of the body — a rise in the level of cholesterol or sugar (for example) can dis-rupt the surface of the vesicle, delivering the drug appropriately.

✔ Endocytosis (shown in the "b" part of Figure 11-8) is the passage of the vesicle into the cell, after which the vesicle breaks down and the drug molecules are delivered.

✔ The "c" part of Figure 11-8 shows an uncommon method: The liposome fuses with the cell wall and delivers its contents into the cell.

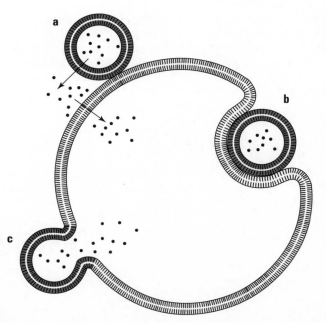

Figure 11-8:
Types
of drug
delivery.

Phospholipids are not the only drug-delivery trick available to us. If we want a bit more durable packaging, we can use polymers to create drug-delivery vesicles. This method allows greater flexibility and structural stability. Trigger molecules can be attached to the outer vesicle wall, allowing it to release in response to contact with a particular cell. Additionally, the bonds between polymer amphiphiles are stronger than those with the phospholipids — minimizing the chance of accidental release. This strength also allows for a controlled release (if desired), and keeps the vesicles stable prior to injection.

Drug delivery at the nano scale has two important elements: the ability to target specific cells and controlled release. Liposomes provide both, so they're sure to be an important part of nanotechnological drug development. They offer three particular advantages:

✔ They have already provided a useful vehicle for successful drug delivery. For example, they have been used to deliver doxorubicin and anthacy-cline, both drugs that treat cancer. Diabetics have also taken advantage of the trigger-release mechanism of liposomes — when glucose levels are high, the vesicles release insulin.

✔ They have been used and proposed for gene therapy. As gene-therapy tools they can be set to encapsulate and deliver segments of DNA and RNA directly into cells to replace specific faulty genes. This technique promises to treat a variety of diseases.

✔ They penetrate the skin, delivering molecules to the deeper layers of cells. This is of great importance to the cosmetics industry for skin creams, sunscreens, vitamins, and even tanning agents.

Unfortunately, the body can quickly identify the liposomes as foreign intruders and destroy the delivery vehicle. In response, researchers have developed "stealth liposomes," coating the liposomes with large, neutral polymers. These barriers interact weakly with the antibodies that attack foreign substances, allowing them to remain in the body for longer periods of time — very important if the liposome is to deliver a time-release drug.

Stepping it up with C_{60}

One other handy chemical tool is a buckyball (that soccer-ball-shaped molecule we discuss at length in Chapter 4). It provides the support structure to an amphiphile, which in turn can be used to create drug-delivering micelles. Figure 11-9 shows the C_{60} buckyball — the molecule made of 60 carbon atoms — surrounded by attached side chains. The top half of the structure acts as the hydrophilic head; the bottom half acts as the hydrophobic tail.

Brother, can you spare a dendrimer?

A *dendrimer* is an artificial, polymer-based molecule that resembles a foam ball with tree sprigs shooting out of it in every direction. A typical dendrimer is the size of a typical protein (3 nanometers across), held together with strong chemical bonds, but also containing a great number of voids — in effect, pockets for carrying payload. The voids make up a great amount of surface area, and can be tailored to different sizes, allowing great control and flexibility in their applications. One such use is to transport DNA into cells for gene therapy — an alternative to genetically modified viruses. Viruses are troublesome as delivery vehicles for therapeutic DNA; the body hates foreign substances, and even a benign virus can trigger severe immune reactions. Dendrimers, on the other hand, covertly bypass the immune defenses and slip DNA into target cells.

Dendrimers may also be used for drug delivery. Given their ability to be controlled, they may be designed to swell and liberate their contents only when the appropriate trigger molecules are present. Additionally, they may be used for diagnostics — they may deposit a type of DNA that makes a diseased cell glow. (More on diagnostics can be found in Chapter 10.)

The interesting development of this is that these C_{60}-based amphiphiles form vesicles of different sizes, depending on what pH levels are present. It's also known that sites of infection have pH levels different from those in the rest of the body — and scientists can tune the amphiphile to respond to specific pH levels. As you can imagine, using C_{60} vesicles as a drug-delivery system looks very promising — the vesicles go throughout the body and release a drug at all sites of infection — those sites with a specific pH level — and nowhere else. Figure 11-10 shows a vesicle utilizing C_{60} amphiphiles that are encapsulating a drug. (Keep in mind that even though the image looks circular, its structure is spherical.)

C_{60} can be used as a "molecular pincushion" attaching drugs and protein detectors to its shell. Fundamentally, C_{60} is nontoxic and can be modified rapidly and easily. Imagine modifying the C_{60} molecule to disrupt the AIDS virus's ability to reproduce. Since the AIDS virus changes and adapts, the ability to modify C_{60} drugs quickly in response to changes in the AIDS virus is a key to longer-term effectiveness.

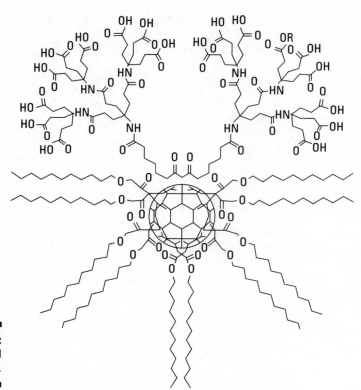

Figure 11-9:
A buckyball
amphiphile.

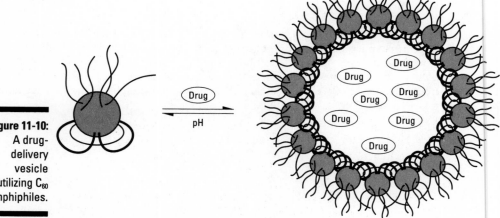

Figure 11-10:
A drug-
delivery
vesicle
utilizing C_{60}
amphiphiles.

Cooking Cancer with Nanoshells

The body is constantly replacing old cells with new ones; the old, damaged cells are deliberately "killed" in a process called *apoptosis*. Sometimes, however, mutations occur so that some new cells form when the body doesn't need them and old cells don't die when they should — which, by the way, is a basic definition of cancer. Cancer cells bypass apoptosis and form a mass of tissue called a *tumor*. (See Figure 11-11 for a graphical depiction of normal cell division versus cancer mutation.)

As it happens, the National Cancer Institute (NCI) has identified four areas of the body as the locations of the most common forms of cancer: prostate gland, breast, lung, and colon. Not surprisingly, the amount of spending on cancer research is particularly high, especially for these four types of cancer, as Table 11-1 clearly shows. The NCI's total budget for cancer research back in 2003 was set at $4.6 billion, so you can see the motivation to find alternative (more effective and less expensive) ways to combat cancer.

Imagine having a way to deliver the knockout punch directly to those four areas. No sooner said . . .

Enter *nanoshells* — a unique, noninvasive way to detect and obliterate cancer cells.

Table 11-1	Cancer Research Spending		
Type of Cancer	*2003 (Actual, in Millions)*	*2004 (Estimated, in Millions)*	*2005 (Estimated, in Millions)*
Prostate	$305.20	$320.50	$337.00
Breast	$548.60	$578.00	$605.00
Lung	$273.40	$288.50	$297.00
Colon	$261.60	$272.00	$279.00

Nanospectra Biosciences, Inc., is a company now capitalizing on the nanoparticle's potential to eradiate cancer — and they've come up with their own approach: They start off with a very small glass bead and coat it with gold. Here's why:

- ✔ **Gold, an inert metal, can absorb quite a bit of light.** Its optical absorption rate varies depending upon how thick a layer of gold is applied.

- ✔ **Gold happens to be biocompatible with the human body.** It can stay in there without corroding or otherwise reacting, as any dentist can tell you.

This means you can vary the size of the glass bead and the thickness of the gold layer to scatter or absorb different wavelengths of light. Figure 11-12 illustrates the scattering wavelength with a 60nm core and various shell thicknesses.

When dealing with medical applications, we're interested in the near-infrared wavelength (650–1050nm) because this is the wavelength that transmits well through biological tissue.

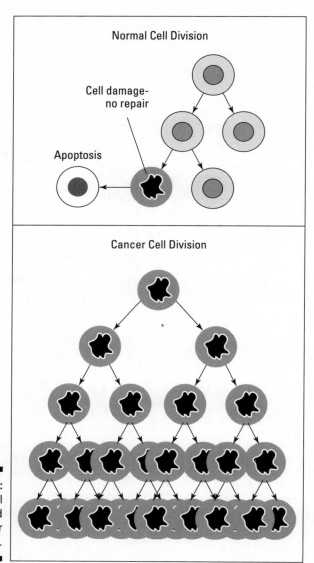

Figure 11-11: Normal cell division and cancer mutation.

"Why does the near-infrared transmit through biological tissue?" We're glad you asked. The body is mostly made up of water. Our goal is to find the best spectral region for optical imaging. This region, as it turns out, is between 800 and 1300nm — and is given the special title of *water window*. We are able to cater our nanoshells to either absorb or scatter within this range for optimal effectiveness through human tissue.

Nanoshells can be used to "cook" cancer cells (absorption), but they can also be used for imaging (scattering). That's a process of absorption and scattering. A handy example is the color of the shirt you wear. If you wear a black shirt during a summer day, you tend to feel warmer because the black color absorbs the light; if you wear a white shirt on the same summer day, you feel cooler because the white color reflects the light. In much the same way, nanoshells can absorb light and give off heat (like that trendy black shirt of yours), but they can also reflect (like the white shirt), allowing a camera to pick up the scattered light.

Figure 11-12: Tuning the absorption and scattering capability of nanoshells.

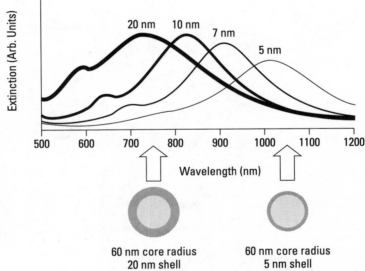

Reprinted with permission from: "Nanoshell-Enabled Photonics-Based Imaging and Therapy of Cancer," Technology in Cancer Research and Treatment, Vol. 3, Num. 1, Page 33, 2004, Adenine Press, Inc. (www.tcrt.org).

But let's get back to the thrilling story of Nanospectra Biosciences, Inc. They create their nanoshells by first growing perfect silica (glass) cores and then coating the cores with a special amine. (*Amines* are organic compounds that can be used as "attachment" points for structures.) Recall that in Figure 11-9 (in the previous section), C_{60} had certain chains hanging from it — chains that changed the amphiphiles' properties and the way they reacted with various pH levels. The process of attaching these chains is called *functionalization* — attaching groups of molecules with a specific function in mind. Using C_{60} as the support structure and functionalizing the side chains gives us a unique amphiphile capable of doing the good stuff we want done. In this case, the amines are functionalized to the glass core — which will act like glue for our next step.

As a functional molecular group, the amine works best for attaching (gluing) gold particles to the glass core. After the amines are attached to the core, the coated silica particles are suspended in a bath of smaller gold particles (less than 2 nm wide). Figure 11-13 shows some gold particles attaching to the amines. A further reaction involving additional chemicals causes more gold particles to attach, creating a shell roughly 10 nm thick. That shell can best absorb light that has a wavelength of 810 nm. Figure 11-14 illustrates the final nanoshell product; Figure 11-15 shows actual images of the silica core surrounded by varying thicknesses of gold particles.

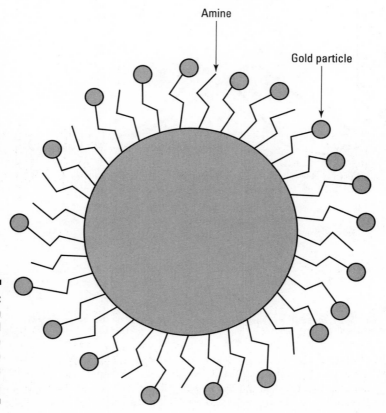

Figure 11-13:
Amines with some gold particles are attached to the glass core.

Figure 11-14:
The final nanoshell with gold-coated core. The thicker gold coating is a result of a progressive increase in the number of gold particles attaching to the shell.

Figure 11-15:
TEM images of gold particles attaching to the outside of a silica nanoparticle core. The gold is biocompatible with the human body.

— 20 nm

Reprinted with permission from: "Nanoshell-Enabled Photonics-Based Imaging and Therapy of Cancer," Technology in Cancer Research and Treatment, Vol. 3, Num. 1, Page 33, 2004, Adenine Press, Inc. (www.tcrt.org).

So now we have a gold-plated nanoshell, not currently doing anything of importance — until we prepare it for medical applications by attaching antibodies to the outside. Why antibodies? Well, antibodies are the body's way of detecting and flagging the presence of foreign substances — a fact we've known about for quite a while. It turns out, however, scientists can now use known cancer cells to create mass-produced, protein-based antibodies in the lab. These antibodies can then be attached to the outside of our gold-plated nanoshells which, when they're injected into the body, then attach themselves to these specific cancer cells — and only to cancer cells.

The search process isn't instantaneous — it takes a few hours for the nanoshells to circulate through the body. Figure 11-16 is a graphical representation of a nanoshell's antibodies attaching to the surface of a cancer cell. Figure 11-17 shows cancer cells first illuminated with laser light showing no change, then with attached nanoshells, and finally illuminated with a laser activating the nanoshells. This last image shows the death of cancer cells at the circular laser point. This experiment was demonstrated *in vitro* — Latin for "within glass," meaning outside the body and in a Petri dish.

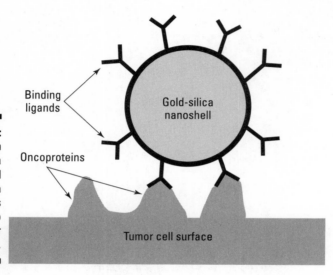

Figure 11-16: Cartoon image of a nanoshell with antibodies attaching to the tumor cell surface.

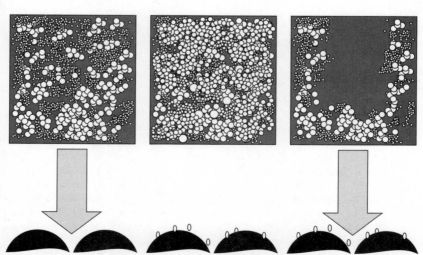

Figure 11-17: Cancer unaffected by laser treatment alone, nanoshells attached to the cancer cells, and the death of cancer cells with the nanoshell/laser treatment.

Reprinted with permission from: "Nanoshell-Enabled Photonics-Based Imaging and Therapy of Cancer," Technology in Cancer Research and Treatment, Vol. 3, Num. 1, Page 33, 2004, Adenine Press, Inc. (www.tcrt.org).

In vitro is a nice start, but Nanospectra Biosciences, Inc., knew that at some point they'd have to go *in vivo* — within the living. In subsequent experiments, live mice were injected with tumor cells. When the tumors reached a particular size, the nanoshells were injected. After six hours, an 808nm laser was used for three minutes to activate the nanoshells. The chart in Figure 11-18 shows the tumor size at Day 0 (treatment day) and Day 10. The nanoshell therapy for seven cases shows no sign of the tumor. For those treated only with the laser (without nanoshells, eight cases) and for those left untreated as an experimental control (nine cases), the tumors continued to grow. The graph in Figure 11-19 shows that all mice with the nanoshell treatment lived the entire 60 days; those treated only with the laser (or without treatment) were euthanized by Day 19, or when the tumor grew to more than 5 percent of their body weight. Yes, it's grim — but cancer is grim business, and nanoshell treatment offers real hope.

Figure 11-18: The tumor sizes for Day 0 and Day 10 of this experiment: By Day 10, the tumor is gone for the nanoshell-treated mice.

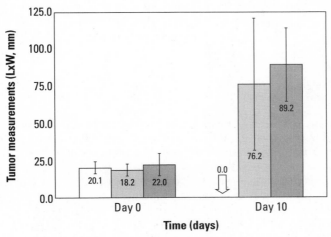

In fact, this result shows tremendous promise — a therapy that can actually destroy cancer cells with minimal invasiveness. This process offers the greatest initial hope for breast and prostate cancers — which appear predominantly at the body's surface, where laser light can be most effective.

A further advantage to nanoshell treatment is detection: Nanoshells can be configured to scatter light as well as absorb it. This scattering can be used to create an image of where the nanoshells congregate. A possible future treatment could have the cancer patient visit the doctor periodically to be injected with special nanoshells coated with antibodies that would search for various types of cancer. Then electronic scanning would find the nanoshell congregations and allow pinpoint targeting of the laser — which would then cook the tumors — all in one afternoon. Human trials for this technique will begin within the next few years.

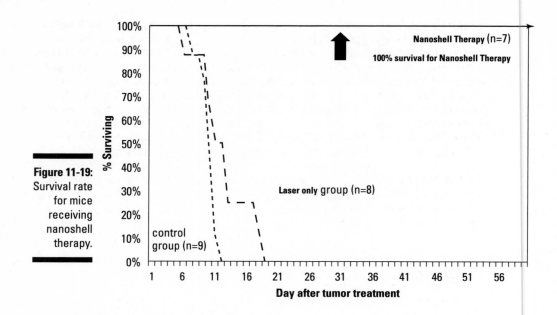

Biomimetics

If imitation is the sincerest form of flattery, then technology has been flattering nature for centuries. *Biomimetics* is the science of mimicking biology — and no, we're not talking about imitating the mating call of the common loon (not just now, anyway). Nature has a few billion years' lead over humans in complex bioengineering — and we've figured out that nature can teach us a thing or two about functional design. Humans, for example, have studied and synthesized rubber and spider silk, both incredibly important natural materials. Synthetic rubber helped us win World War II — military items such as engine belts and tires relied upon synthetic rubber after the Axis powers cut off our natural rubber supplies from Malaysia. And where would we be if we hadn't started mimicking the shape of a bird's wing and incorporating it into airplane designs? Still walking, I assume.

The biomimetic goal is to figure out how nature does what it does, and then develop techniques that beat nature at its own game. Humanity is like an apprentice who learns from — and then supersedes — the master. A telling example is the medical advances that have improved oxygen delivery and artery expansion (described in the next sections). Within the human body, biomimetics takes on a new goal: to encourage the body to help repair and regenerate itself. Two current examples are joint-replacement therapy and tissue scaffolding, both of which combine biology and chemistry — and manipulate biochemical processes at the nano scale.

Synthetic materials do not make good implants — they wear out relatively fast, and can't regenerate themselves the way bone can. They also have an effect on the body — and are affected by the body — in ways that limit their usefulness. For example, the average lifetime of orthopedic implants (including replacement hips and knees) is about 15 years — not a happy prospect for a 20-year-old, who may need multiple replacements over a lifetime. The implants must be replaced because the patient's bone *resorbs* — disappears from around the implant — which loosens the implant and makes it easier to break. Similarly, vascular grafts using artificial blood vessels have remained only 25-percent functional after they've been in place for five years.

So what's the problem here? In a word, size: Synthetic material made of particles bigger than 1 micron won't let natural tissue regenerate, which shortens the life of the implanted device. One answer is nano-size materials — made of particles with dimensions smaller than 100nm.

Improving oxygen delivery

The trouble with breathing is that it's only the beginning. After you inhale, the oxygen has to get to your internal tissues via your bloodstream. Tissue damage results from poor blood flow and inadequate oxygen supply. In fact, the victims of heart attack and stroke suffer greater damage from a lack of oxygen than from the initial attack. Blood transfusions have saved lives but have three main drawbacks:

- ✔ **Blood type:** The body rejects blood that doesn't match what it makes for itself.
- ✔ **Viral contamination:** Viruses, especially the sneaky ones such as HIV, are always a risk.
- ✔ **Perish ability:** Natural blood has a short shelf life, which is one reason it's often in short supply.

Even with good matching of blood type and effective screening of viruses, creating an artificial blood remains attractive: Imagine a plentiful supply of new, virus-free blood that could match any blood type — and even *work better* than natural blood.

The ideal blood substitute has been hard to come up with so far. One approach uses purified *hemoglobin,* the oxygen-carrying protein in blood cells. Hemoglobin doesn't have a specific blood type, can be sterilized, and may be stored for more than a year. Unfortunately, it was quickly discovered that using this purified hemoglobin by itself as a blood substitute proved highly toxic to human kidneys. The purified hemoglobin broke down into smaller units than normal hemoglobin and quickly accumulated in the kidneys — to toxic concentrations — which put severe limits on how (and if) it could be used.

This is where nanotechnology offers a couple of new and different solutions. Solution # 1 links several hemoglobin molecules into a larger complex, as in Figure 11-20. Modifying hemoglobin — say, by fusing two hemoglobin sub-units into one chain — forms a stable, complex molecule that carries oxygen extremely well — and is too big to be filtered out by the kidneys before the oxygen reaches its goal.

Figure 11-20:
Two sub-units of hemoglobin (pink) combined into one long chain, suitable for use in blood replace-ment.

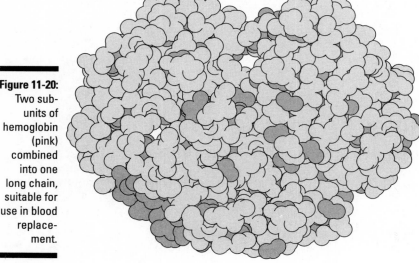

Solution # 2 takes unaltered hemoglobin and encapsulates it in liposomes — essentially the same way liposomes are used for drug delivery. (See the "Special delivery" section, earlier in this chapter, for all the details.) Now, these liposome-based, artificial red blood cells don't provoke an immune response (that's the good news) — but they *are* removed from the blood-stream by the same process that disposes of defective red blood cells (that's the bad news). In the face of this hurdle, scientists have modified the outer shell of the liposome so it resembles a healthy red blood cell. Otherwise the body would see it as broadcasting to the world, "I'm defective, I'm defective. Trash me."

How does this microscopic masquerade work? By modifying the lipids that make up the liposome so they resemble the sugars that coat the red blood cell. This coating of sugars varies by blood type; if the liposome looks like a cell of the correct blood type, the body lets it go about its business.

Solution # 3 makes use of biodegradable polymers that have been fashioned into nanocapsules to transport hemoglobin. Such polymers (for example, polylactide and polyglycolactide) degrade in the body, producing water and carbon dioxide as harmless byproducts. Such nanocapsules have some advantages over our liposome transport mechanism — these are stronger, more porous, and can carry more hemoglobin.

They can also carry the stuff needed to keep the hemoglobin from generating toxic chemicals (such as the oxygen radicals in this case). Natural red blood cells contain enzymes that destroy oxygen radicals — and since there's plenty of room in polymer nanocapsules, the detoxifying enzymes can go along with the hemoglobin. Result: a safe, nontoxic, effective blood substitute.

Now for Solution # 4, which admittedly sounds a bit farfetched — but, hey, what chapter of this book would be complete without something that sounds a little like science fiction? In the 1989 movie *The Abyss,* director James Cameron introduced a Fluid Breathing System (based on actual, successful experiments with mice) that pumps an oxygen-rich fluid into the lungs so a diver can breathe despite the great pressure at extreme depths — and avoid "the bends" on the way to the surface. The concept is kind of like a baby breathing in the womb, with amniotic fluid instead of air.

Thinking nanotechnologically, you can take this concept a step further and introduce tiny spheres filled with high-pressure oxygen into the bloodstream. These tiny spheres, known as *respirocytes,* could be made to release oxygen at a constant rate. In effect, a respirocyte does the work of a red blood cell — but more efficiently, and with a bit more of an exotic mechanical flair. Figure 11-21 shows how such a respirocyte could look: A rotary assembly around the respirocyte exchanges oxygen, carbon dioxide, and water — and provides locomotion through the body. Figure 11-22 is an artist's conception of respirocytes in action.

Expanding an artery from the inside

Clogged arteries are not just about inconvenient traffic jams. Annually, $53 billion is spent to treat strokes — and in 80 percent of these strokes, a blood clot clogs a blood vessel to the brain. According to the American Stroke Association, strokes are the third-leading cause of death among Americans (heart disease and cancer are the top two).

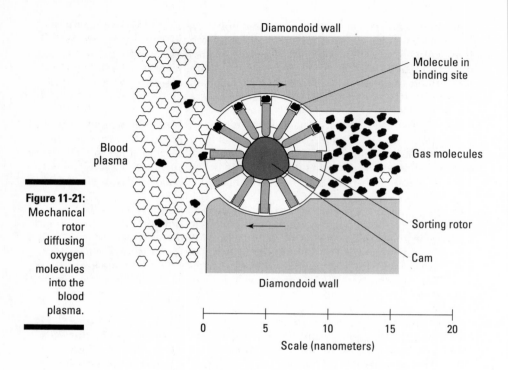

Diamondoid wall

Molecule in
binding site

Blood
plasma

Gas molecules

Sorting rotor

Cam

Diamondoid wall

Figure 11-21:
Mechanical
rotor
diffusing
oxygen
molecules
into the
blood
plasma.

Scale (nanometers)

0   5   10   15   20

Figure 11-22:
Artist's
conception
of respiro-
cytes at
different
angles.
(Looks a bit
like the
floating
target-
practice
device in
Star Wars,
doesn't it?)

One (more or less traditional) way to combat clogged vessels is to thread a tiny corkscrew through an artery to clear the blockage and restore blood flow. This tiny corkscrew, called the Merci Retriever, has recently been approved by the FDA and has undergone successful tests. There are some limitations to the Merci Retriever, however: It must be used within three hours of suffering a stroke, the clot must be visibly located, and the blockage must be accessible by inserting the device into a blood vessel. These problems, as daunting as they are, can be addressed with nanotechnology:

 ✔ **Find the problem:** Properly and accurately identify the location of the blood clot. What's needed here is good imaging (as described in Chapter 10) that can find a blood clot.

 ✔ **Get the device to the site:** Access the too-small-to-reach places — small enough to be easily threaded into an artery.

 ✔ **Do it in a timely manner:** That means — among other things — avoiding the need for extensive preparation in an operating room. If we could move the imaging equipment and Merci Retriever to the ambulance, we could treat patients at the scene — increasing their odds for successful recovery.

One further question: When you've created and used a mechanism that can physically destroy a clot, how do you prevent the problem from happening again? In the past, we've repaired ailing arteries by using tissue grafts from patients themselves, thus ensuring no negative immune response. However, this is not always possible because the *harvesting sites* (the places where we get the patient's tissue graft) may become infected — and tissue grafts in general have limited shelf life. These factors have led to a great demand for an artificial material that has the appropriate mechanical properties needed for arteries: tough but flexible, adaptable to surrounding tissues, and not subject to buckling, kinking, or allergic reactions.

One solution is to work with the blood vessel you're stuck with — and simply improve upon it by introducing a *stent* — an expandable wire mesh used to keep the vessel open. (Think of it as a tube made of chicken wire.) Figure 11-23 shows the process of introducing the stent. The leftmost panel in Figure 11-23 shows a balloon placed at the area of blockage and inflated to make room for the stent. The balloon is then deflated and the stent is placed in position (second panel). The balloon is then inflated once more to expand the stent, which stays expanded as shown in the third panel. Finally, the balloon is deflated and removed, leaving the stent in place, allowing blood to flow naturally, as shown in the final panel.

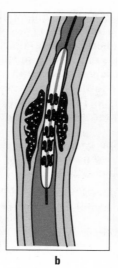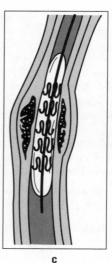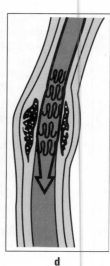

Figure 11-23:
Introducing
a stent into
a blood
vessel.

a   b   c   d

This approach is all fine and dandy, but it ain't nanotechnology. The stent we describe here is as traditional in its engineering as a wooden tongue depressor — not exactly cutting-edge. And traditional stents are not problem-free. Sometimes they actually cause clogging — the problem the stent was supposed to correct in the first place. The result is as serious as its name sounds: *Thrombosis* is the deposition of coagulated blood material, formed from a two-step process: coagulation of some of the proteins in blood and the depositing of cells called platelets. Once the blood coagulates, the platelets come in and stick to the forming clot, which in turn coagulates more blood, which in turn leads to deposits of platelets, and so on until blood flow is blocked. Natural blood vessels are supposed to combat clogging by lining the arteries with *endothelial* cells, which produce heparin (to prevent coagulation) and prostacyclin (to prevent platelet deposits). Traditional stents have no way of working with endothelial cells, so they have no effective way of staving off thrombosis.

Advanced Bio Prosthetic Surfaces (ABPS), a company based in San Antonio, Texas, feels it may have come up with a solution to our little stent problem. They've taken as one of their research and development goals combining technology from the semiconductor industry with the medical industry, and it just so happens that they've lighted upon the idea of using shape memory alloys such as Nitinol (ably discussed by yours truly in Chapter 5) as material for stents. Using a shape-memory alloy is a great idea because it remains flexible but springs back to maintaining its shape.

Now, the use of this alloy is not new in stents, but ABPS has come up with a new process for creating the stent using this metal. Old processes involved macro-scale manufacturing — which introduced micro-scale cracks on the surface of the stent. These cracks, as small as they were, could irritate the artery, causing complications such as infection or blood-flow restriction. At ABPS, the stents are created by *sputter deposition* — a method of creating a thin film of metal by "sputtering" fine particles of Nitinol onto a surface. This allows layer-by-layer, atom-by-atom fabrication of materials, virtually free from contaminants and impurities.

The old method of making a stent causes the surface to be plagued with splinters and other imperfections that could react with the body. The new method involves covering the stent with a nanoporous coating — not only making the stent less likely to trigger inflammation, but also thinner, stronger, and more flexible. As an added bonus, it promotes tissue incorporation and *endothelialization* — the creation of the endothelial cells that prevent clotting.

Replacing joints with better stuff

Thus far, nanotechnology offers almost magical transformations of materials, strengthening our current versions of standard steel, glass, and plastic into something revolutionary. (Don't believe us? Check out Chapter 5.) When applied to biological materials, nanotech will accomplish similar dramatic changes. To do so, however, it will have to compensate for biomaterials' dynamic (ever-changing) construction. Not sure what's so dynamic about biomaterials? Read on to find out.

Bones are actually a sort of process: At the cellular level, they undergo constant repair and rebuilding as they fight gravity, allowing our body to effectively adapt to our environment. (That's pretty dynamic, right?) What makes bone so versatile is its basic building material: hydroxyapatite. Osteoblasts (bone-forming cells) secrete this stuff within a matrix of collagen (the main protein of connective tissue, which is why it shows up so often in beauty products). The body continuously reabsorbs the hydroxyapatite crystals and re-deposits new material in their place. What that process does is adjust the bone's thickness in response to changes in the body's distribution of weight. Figure 11-24 shows an excellent top-down hierarchy of strength in the materials that make up bone — comparing actual bone cells to collagen fibers and to individual collagen molecules (each a mere 3nm in length).

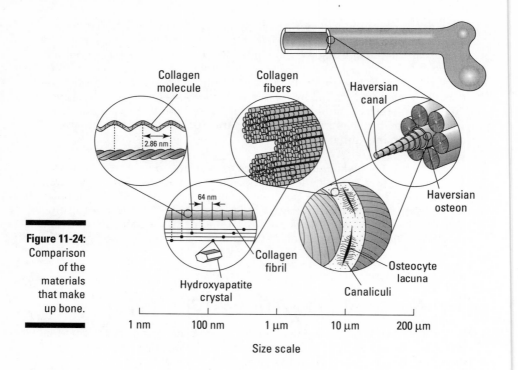

Figure 11-24:
Comparison
of the
materials
that make
up bone.

All that specialized strength gives human bioengineers quite a challenge. Implants are typically made of highly inert material — stuff designed to interact as little as possible with the body in order to minimize the implant's interference with the normal growth and function of surrounding cells. With traditional implants — like the ones presently used for replacement hips — the bone cells eventually disappear from around the implant, loosening it and making it prone to breaking. Why is the bone being reabsorbed? Well, the body figures the bone isn't needed. Since the implant is now bearing the greatest amount of load, the body responds by "pulling back" on bone production in that area. Just as the osteoblasts are stimulated into generating more bone when they're confronted by more stresses, they can also be told to back off when (falsely) the burden seems to be less. Result: Bone generation falls off, and then the body starts to reabsorb existing bone. The problem is how to keep the bone in the business of rebuilding itself.

Looking specifically at hip implants, another source of contention is the plastic that is used in the socket portion. A traditional hip implant has a metal ball and a plastic socket, as shown in Figure 11-25. As the metal rubs against the plastic socket, plastic particles break off, and the body responds to the foreign fragments with inflammation, resulting in bone loss at the femur. Researchers at the University of Tokyo are developing a biocompatible polymer joint that resembles phospholipids (the cellular membranes we discuss in the "Delivering a New Drug the Nanotech Way" section, earlier in this chapter). Using such a polymer joint fools the body into not attacking the residual flakes; no inflammatory response means no weakening of the bone. This is an example of designing implants that interact with the body on its own terms — an improvement over the old school of thought that largely ignored the interaction between implant and body.

Artificial Hip Joint

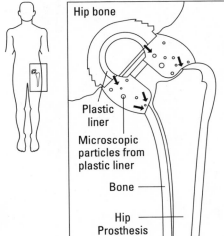
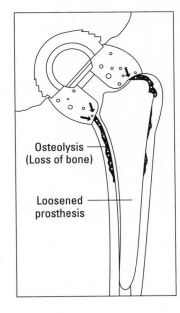

Figure 11-25:
Hip implant
cross
section.

Bioactive materials have the ability to interact with living tissue — and are the most promising approach to ensuring a strong, long-lasting adhesive interface between the implant and the surrounding tissue. Already some bioactive materials have done such a good job at mimicking natural bone that tissues have chemically bonded to their surfaces, successfully persuading the body that the implant is not really a foreign substance.

One bioactive material already in clinical use — and showing tremendous promise — is being developed by Doxa, a Swedish biomaterials company. They have come up with a process that involves injecting the affected joint with a soft material that then hardens at the molecular level, bonding a damaged bone together. More specifically, the material reacts to body fluids to form *apatite,* the body's own ceramic. The resulting material is better than existing artificial-bone material; it's as strong and hard as actual bone. Such a material is considered a "bioactive ceramic" — the body activates the material and helps form it into a ceramic. The approach combines various sophisticated understandings of materials, chemistry, and biology at the molecular level, which is what nanotechnology is all about.

Part V
Investing in Nanotech

The 5th Wave   By Rich Tennant

"Not quite nano-size yet, but we're getting there."

In this part...

So who's doing what out there to move nanotechnology along to where it can deliver on all this promise?

So glad you asked. In this part, we look at three segments of the world that are putting a lot of time, energy, brainpower, and money into nanotech: corporations, government, and higher education. Chapter 12 shows you some companies who are already reaping rewards from nano products on the market today — and others who are heavily invested in following suit with nano-R&D. In Chapter 13, we mosey on over to the world of government, where labs and agencies are working hand-in-hand to ensure that we have the knowledge and workforce to move nanotechnology forward. Finally, in Chapter 14 you go to school and roam around the campus nanotech facilities of several universities who are getting small in a big way.

Chapter 12

Industries Going Small

● ●

In This Chapter

▶ Improving semiconductors with nanotechnology

▶ Healing with nano

▶ Enhancing materials at the nano level

▶ Providing nano raw materials

▶ Designing software for nano-modeling

▶ Building testing equipment for nano-processes

▶ Changing the face of telecommunications

▶ Generating energy with nano

▶ Looking good with nanotech cosmetics

● ●

*A*lthough the big nano emphasis (so to speak) in today's corporate world is in the semiconductor industry — where the reality of making electronics smaller is close at hand — other industries are exploring nanotechnology as well. These include companies making materials such as cloth and rubber stronger as well as those bioengineering companies exploring the uses of nanotechnology in making medical testing and treatment devices more precise.

Here, then, are approximately ten industries and a host of companies that are putting their money where their corporate mouths are when it comes to research and development in nanotechnology.

Semiconductor Types Are Completely into Nano

One of the most obvious applications of nanotechnology is in making computer chips (and other components) tinier and tinier in order to produce smaller electronic devices. Why? Well, people just seem to love small. As sci-fi author Arthur C. Clarke observed, "any sufficiently advanced technology is indistinguishable from magic." When technology makes things tiny, people are intrigued by the voodoo of it all.

Just think about portable phones which shrank from the size of a loaf of bread to the size of a candy bar over the last few decades; people have gobbled up these small electronic devices like — well, candy.

The semiconductor field is a 100-billion-dollar-plus industry, so it's a big factor in today's nanotechnology research. Interestingly, these companies are cooperating in moving the technology forward — because every company will benefit from the research and be able to release new products in their own lines. Organizations such as the Semiconductor Research Corporation (SRC, an arm of the Semiconductor Industry Association) are funding research at universities such as Georgia Tech and Carnegie Mellon — and sharing the results of that research across the industry.

Bell Labs (www.bell-labs.com) has a partnership with the Defense Advanced Research Projects Agency (DARPA) to work on micro-electromechanical systems (MEMS) as a basis for the spatial light modulators (SLMs) used in maskless lithography. Lithography has to do with printing patterns on semiconductors. Traditionally, lithography is done with costly pattern "masks" that can only be used once each. Bell Labs technology does away with those masks, saving money and making it possible to build smaller circuits.

Other leading lights in this sector are Intel (www.intel.com), HP (www.hp.com), and IBM (www.ibm.com) because of their interest in the manufacture of computer chips and parts.

- ✔ Intel started exploring nanotechnology in 2000 — and began *shipping* sub-100-nanometer transistors for its chips in 2003. Their future focus is on silicon-based nanotechnology; they project that a new quantum-well transistor they're working on could use a tenth of the power while still delivering current levels of performance.

- ✔ HP is working on getting rid of transistors and replacing them with molecular-scale switches known as *crossbar latches*. These switches would be initially used in computer memory and could eventually be used to process logic operations.

- ✔ IBM, one of the biggest spenders, has said that self-assembly techniques for building products such as a nanocrystal FLASH memory device will be used in pilot projects within the next two to four years.

Other companies working in the semiconductor nanotechnology realm include

- ✔ Seiko (www.seiko.com)
- ✔ FormFactor, Inc. (www.formfactor.com)

- ✔ PSI Technologies (www.psitechnologies.com)
- ✔ Texas Instruments (www.ti.com)
- ✔ Hitachi (www.hitachi.com)
- ✔ Motorola (www.motorola.com)
- ✔ Samsung (www.samsung.com)
- ✔ Freescale Semiconductor (www.freescale.com)
- ✔ Lucent Technologies (www.lucent.com)
- ✔ Eastman Kodak (www.kodak.com)

See Chapter 6 for detailed info on what's being done in the computer nano-realm.

Mining the Medical Possibilities of Nanotechnology

The potential applications for nanotechnology in medicine are many, but some of these applications need a bit longer development time than you'd find in some other industries. That's due in part to the complex interactions among all those tiny components, as well as tight regulations for new products. In the future, medical tools will be made more inexpensively, diagnosis will become more accurate, and we can expect tiny, inexpensive sensors and implants to provide both automated monitoring and semiautomatic treatment. Medical imaging and in-vivo drug delivery are also being explored.

Nanospectra Biosciences (www.nanospectra.com) is one company pushing the envelope in nano-medical applications today. They have patented nanoshell particles used in noninvasive medical therapies. These nanoparticles can actually be tuned to scatter or absorb light in specific wavelengths. This makes possible tasks such as destroying specific cancer cells with an infrared laser, welding tissues to heal wounds, and something called *photocoagulation* (coagulation with a laser) that cuts off the blood supply to unhealthy cells, slowing down degeneration in people with diseases such as diabetes and cancer.

Hitachi (www.hitachi.com) established Hitachi High-Technologies in 2001 to pursue nanotechnology in a variety of settings, including medical testing in conjunction with Hitachi Medical Corporation and Hitachi Instruments Group. Hitachi is exploring DNA analysis through the use of atomic and molecular devices. Hitachi also recently partnered with Oxford and Cambridge in a joint venture called Nanotech.org. They manufactured the world's tiniest test tube (you could fit about 300 million of them into a grain of sand). What people will do with these test "tubettes" is limited only by the human imagination. (For now you'll just have to imagine what they look like.)

Dow Chemical (www.dow.com), through its division DowPharma, is actively involved in *drug solubilization* technology, working on altering nanoparticle size and properties to make drugs easier for the body to absorb. It's a key aspect of effectively delivering small-molecule drugs to where they can do the most good.

Other companies dealing with medicine, medical testing, and drug-related nanotechnology are

- ✔ Merck (www.merck.com)
- ✔ Abbot Laboratories (http://abbott.com)
- ✔ Beckman (www.beckman.com)
- ✔ Nanoprobes (www.nanoprobes.com)
- ✔ American Pharmaceutical Partners, Inc. (www.appdrugs.com)

Making Better Materials from Tires to Clothing

Many companies are already using nanotechnology to make better materials — from the rubber used for tires to stain-repellent fabrics for clothing.

The name Zyvex pops up whenever you search for nanotechnology companies online. Founded in 1997, this company's slogan is "Providing Nanotechnology Solutions — Today(tm)" — and so they are. Zyvex has a line of nano-additives and enhanced carbon nanotubes that add strength to composite materials. Using these they can improve thermal, electrical, and mechanical properties of various materials. Zyvex counts among its customers aerospace, electronics, telecommunication, and healthcare companies.

Are you a nanotech investor? In this era of nano speculation, Zyvex is one of the few companies with strong funding and solid business practices that is actually making money off of nano products and services. They're a private company, but if they ever go public, call your broker!

In clothing manufacture, Gore-Tex (www.gore-tex.com) offers a jacket made of a waterproof polymer membrane with embedded nano-size carbon particles that resist taking on a static charge. Such anti-static protection could be used, for example, in firefighters' suits to help them avoid static electricity that could spark fire in a hazardous situation. Nano-Tex (www.nano-tex.com) uses nanotechnology to produce Nano-Care stain-resistant clothing. Nano-Tex won't give away how they incorporate their spill-resistant technology, but it's likely that they coat the fibers with a polymer. Because the fibers are coated and then woven into fabric, the cloth maintains its stain resistance even after many washes — and the fabric maintains its soft texture and feel. They license the technology to others, including clothing giants The Gap, Nike, and Old Navy. Many American and European textile companies are counting on their high-tech nano edge to survive as they compete with countries that can provide lower-cost labor.

Nano-based materials run the gamut, even appearing in the sports arena. Babolat (www.babolat.com), a French maker of tennis rackets, for instance, produces a racket that uses carbon nanotubes for higher torsion in the racket and flexibility of the racket strings. Other companies are developing tennis and golf balls that fly straighter because of the nanotechnology inside.

InMat (www.inmat.com) is using nanotechnology for everything from packaging that keeps food fresher to tires that hold air longer (and are lighter in weight for improved mileage). InMat is also into sports, with soccer balls and tennis balls that last longer and work better. InMat achieves many of its successes thanks to its Nanolok aqueous coatings (nanoparticles dispersed in silicate and suspended in a polymer).

General Electric (www.geglobalresearch.com) is focused on material performance. They are working on nanotubes and nanorods, nano-ceramics, nano-structured metal alloys, and self-assembled block polymers. Given the breadth of GE products (everything from ovens to aircraft engines and diagnostic-imaging scanners), there are endless applications of improved material performance for this electronics giant.

Other companies exploring how to make materials stronger with nanotechnology include

- ✔ Altair Nanotechnologies, Inc. (www.altairint.com)
- ✔ Celanese (www.celanese.com)
- ✔ Materials Modification, Inc. (www.matmod.com)

Making Nanotech Materials for Others

Several companies are producing the raw materials of nanotechnology for use by others. Carbon Nanotechnologies Incorporation (www.cnanotech.com) produces various grades of nanotubes, as well as sheets of graphite for use by a variety of industries.

QuantumDot (www.qdots.com) does work in quantum dots, tiny crystals that emit light of various colors. These are a big improvement over fluorescent dyes traditionally used in cell research (which light up only briefly and tend to bleed into each other). Quantum dots are being used by various companies in genotyping (tagging the various genes in the genome with what is essentially a color-coded bar code), analysis of white blood cells for HIV patients, and even blood fingerprinting.

LG Electronics (www.lgelectronics.co.uk) makes a product called Nano Carbon Ball, a hollow sphere that they coat with a very thin layer of nanoporous carbon. These balls soak up odors ten times better than other carbon materials. The first use of this material is in a refrigerator used to ferment kimchi in Korean markets.

Other suppliers of nanotechnology materials include

- nGimat (www.microcoating.com)
- Nanomaterials (www.nanomaterialscompany.com)
- NanoDynamics (www.nanodynamics.com)
- Zyvex (www.zyvex.com)
- FEI Company (www.feicompany.com)

Designing for Small with Software

Coventor is a company that produces software products to help companies develop products based on Micro-Electro-Mechanical Systems (MEMS), semiconductors, and microfluidics. They tout themselves as having an installed base of users larger than those of their top ten competitors combined. Check out their Web site (www.coventor.com) for a free software evaluation offer if you want to experience their products firsthand.

Coventor developed (in a partnership with Zyvex) MEMulator — a software program used for emulating and viewing nanotechnology processes. In addition, the aptly named CoventorWare, according to the company, performs the following futuristic magic: "enables schematic-based behavioral modeling and detailed 3-D multi-physics numerical analysis." Now, there's a mouthful of fancy words — but they just mean that CoventorWare is a suite of programs for producing MEMS designs; it can be used to create nanotechnology models, simulate how these models work, and analyze the results.

Accelrys (www.accelrys.com) manufactures modeling and simulation software for use in nanotechnology. Their software has been used to design drugs, model modified polymers, and to try out ideas in the materials sciences. Their products use quantum mechanics, molecular dynamics, and simulations to predict properties, understand mechanisms, and interpret data such as measurements in nanotubes.

Other companies that offer software for use in nanotechnology research and testing are

- ✔ NanoTitan (www.nanotitan.com)
- ✔ General Nanotechnology (www.genano.com)

Testing Things

Plenty of companies are making equipment for testing and manipulation of nano-size materials. Good thing, too; making more and more stuff tinier and tinier is probably a growth industry (so to speak).

Veeco Instruments (www.veeco.com) builds atomic force microscopes and other testing equipment for research. Environmentally-controlled equipment can test reactions of various sample materials in different environments. Some of the scanners Veeco manufactures can be used for high-resolution imaging of nano-size materials.

Zyvex builds a nano-manipulator system that mounts on a scanning electron microscope (SEM) to assist in the testing of items such as nanotubes. Another version of their manipulator is designed for use in the Life Sciences as a tool for characterizing biological materials.

Obducat AB (www.obducat.com) is a Swedish company that makes nanotech tools, including scanning electron microscopes and nano-scale lithography equipment used in making computer chips. Their nano-scale lithography equipment is being used at several research institutions, as well as by a number of major companies, including General Electric and Intel.

Other companies involved in producing various types of testing equipment are

- ✔ DuPont (www.dupont.com)
- ✔ Nanometrics (www.nanometrics.com)
- ✔ MTS Systems Corporation (www.mtssystems.com)
- ✔ LabNow Inc. (www.labnow.com)
- ✔ Applied Nanofluorescence (www.appliednanofluorescence.com)
- ✔ Zygo (www.zygo.com)

Technology That's Changing Telecommunications

Improving the alignment of fiber-optic connections is one of many ways to apply nanotechnology to telecommunications. This improved accuracy comes about by aligning the optical fibers — improving the medium via nano-manipulation — which will improve efficiency of transmission.

Bell Labs (www.bell-labs.com), a division of Lucent Technologies, sponsors The New Jersey Nanotechnology Consortium (NJNC), partially supported by the State of New Jersey and partnering with state universities such as Rutgers and The New Jersey Institute of Technology. Bell Lab micro mirror technology improves the modulation of beams used in projection optics. This could be used to improve military and other communications.

Mobile telecommunications are also benefiting from nano. Agilent Technologies (www.agilent.com) is shipping a MEMS device that is helping cellphones get smaller. The MEMS device (called a duplexer) is about 1/20th the size of the part it replaced. Knowles Acoustics (www.knowles acoustics.com) is also tinkering with the use of nanotechnology in producing silicon microphones for cellphones.

Other companies pursuing nanotech advances in fiber optics are

- Corning (www.corning.com)
- Memscap (www.memsrus.com)

Fueling Energy with Nano

Among other things, the Palo Alto–based Nanosys (www.nanosysinc.com) develops nanotechnology-enabled fuel cells used in portable electronics. These fuel cells could power laptop computers, cellphones, or cameras, for example.

In conjunction with companies such as Sharp Electronics, Nanosys is exploring ways for higher-energy density fuel cells to overcome the limitations of traditional batteries, making for longer energy life for portable electronics. They are also working with the United States government on nanotechnology-enabled solar-cell development. A privately held company, Nanosys doesn't stop with energy products. The company holds about 300 patents in nano-technology, and has partnered with companies such as Intel and Dupont to develop products in optical electronics, communications, and defense.

Konarka Technologies (http://konarka.com) produces solar panels with a difference: Their light-activated "power plastic," which contains titanium oxide nanoparticles coated with a dye, is cheaper to produce and more lightweight than solar panels. This material also uses more of the light spectrum than traditional solar cells so it can store energy from any light source, not just the sun. But Konarka doesn't stop at solar panels; they expect to use their "power plastic" in applications such as cases for cellphones and laptops to save you the trouble of plugging them into the wall to recharge their batteries.

Other companies energizing the world with nanotechnology are

- Hydrogen Solar (www.hydrogensolar.com)
- GEMZ Corporation (www.gemzcorp.net)
- Siemens (www.siemens.com)
- Chevron Texaco (www.chevrontexaco.com)

Making Up with Nanotechnology

Proctor & Gamble (www.pg.com) uses polymer and nanotechnology in cosmetics. Using this approach, they can improve how make-up adheres to the skin.

Several companies are producing sunscreen lotions based on mineral nanoparticles for better UV protection. Titanium oxide nanoparticles, for example, provide good protection but don't turn powdery white as the sunscreen dries, as has traditionally been the case. L'Oreal (www.loreal.com) has produced an anti-wrinkle cream that uses a polymer capsule to move active agents (such as vitamins) into the skin.

Bionova (www.ibionova.com) is a company producing nano-skin technology that supposedly mimics the human body's use of bionutrients — chemicals that assist in self-healing. Bionova uses short chain polypeptides and glycoproteins that they claim can reduce wrinkles. They produce a line of skin-care products available to the general public. Interestingly, their discoveries came out of their founder's work with trauma patients and his observation of how the human body can act to heal in times of physical trauma. Sometimes the boundary between medicine and cosmetics can be pretty thin.

A few other companies exploring the use of nanotechnology in cosmetics are

- Jafra Cosmetics (www.jafra.com)
- Ayurveda India (www.ayurvedaindia.org)

Chapter 13

Countries Investing In a Nano Future

*T*he governments of the world aren't immune to nano-fever. Many countries understand the great potential value of nanotechnology in various areas, including defense (think more powerful explosives and stronger and lighter bulletproof vests, for example) and as an engine for economic growth.

The United States is definitely taking the lead, but Europe, Asia, and Israel are also big players. This chapter gives you an overview of some of the governments with initiatives in nanotechnology, along with the key laboratories in the United States.

Showing Nano-Initiative, U.S. Government Style

Nanotechnology isn't a cheap business, so it makes sense that the richest country in the world is taking it on in a big way. The United States is a hotbed of nanotechnology activity as the U.S. government joins forces with industry and academia to build expensive facilities and labs around the country and get our workforce ready to move nanotech forward.

Two National Nanotechnology groups at work

Everybody knows that when the U.S. government works on anything, many agencies are likely to jump into the fray — which can cause duplication and confusion. Something called the National Nanotechnology Initiative (NNI) is a U.S. government program to coordinate nanotechnology efforts in the various agencies. (You can check out their Web site at www.nano.gov.) A presidential-appointed National Science and Technology Council runs the show at NNI.

The stated goals of the NNI are pretty lofty:

✔ Maintain a world-class research-and-development program aimed at real-izing the full potential of nanotechnology.

✔ Facilitate transfer of new technologies into products for economic growth, jobs, and other public benefit.

✔ Develop educational resources, a skilled workforce, and the supporting infrastructure and tools necessary to advance nanotechnology.

✔ Support responsible development of nanotechnology.

NNI does several things. They fund various research projects and contribute dollars to help create university and government nano-R&D labs. The organization also works to help promote the kind of educated workforce that we'll need to ensure that nanotechnology has a future.

Finally, this group encourages small-to-large businesses to jump on the nanotechnology bandwagon. In fact, a major goal of the NNI is to maximize the return on federal dollars invested in nano-scale R&D. That's why they emphasize communicating with the business community on current research activities, partnership opportunities, and available resources. To that end, the NNI is also a matchmaker of sorts, working behind the scenes to promote cross-disciplinary networks and partnerships — all spreading the gospel of nano.

NNI, however, is not the only game in town. The National Science Board, an advisory body of the National Science Foundation (NSF), has okayed funding for the NNIN — the National Nanotechnology Infrastructure Network (www.nnin.org), a joint venture combining the resources of 13 university programs. These schools will eventually make up an integrated, nationwide system of user facilities. Together they'll support research and education in nano-scale science, engineering, and technology. Members of the National Nanotechnology Infrastructure Network may also get access to government facilities such as the DOE User Facilities and the Naval Research Laboratory's Institute of Nanoscience. We discuss NNIN in more detail in Chapter 14.

A whole host of government agencies

There are no less than 11 U.S. government agencies or departments with research-and-development budgets for nanotechnology. Their budgets and focuses vary, and logically relate to their core missions. For example, NASA is working on using nanomaterials to reduce the weight of spacecraft, while the Environmental Protection Agency is funding studies into the use of nano-materials to clean up pollution — as well as studies to evaluate the potential hazards of nanomaterials themselves.

In case one of these initiatives catches your fancy and you want to learn more about its latest efforts, here's a list of the 11 sponsoring agencies, with their individual Web sites:

- ✔ National Science Foundation (www.nsf.gov/crssprgm/nano)
- ✔ National Institute of Standards and Technology (www.nist.gov/public_affairs/nanotech.htm)
- ✔ NASA (www.ipt.arc.nasa.gov)
- ✔ Department of Energy (www.sc.doe.gov/bes/NNI.htm)
- ✔ Environmental Protection Agency (es.epa.gov/ncer/nano)
- ✔ Department of Defense (www.nanosra.nrl.navy.mil)
- ✔ National Institutes of Health (www.becon.nih.gov/nano.htm)
- ✔ National Institute for Occupational Safety and Health (www.cdc.gov/niosh/topics/nanotech)
- ✔ Department of Justice (www.usdoj.gov)
- ✔ Department of Homeland Security (www.dhs.gov/dhspublic)
- ✔ Agriculture Department (www.usda.gov/wps/portal/usdahome)

NASA and DOD have also funded nanotechnology research labs at various universities. See Chapter 14 to discover more about which universities are benefiting from this type of funding.

Nano in the Lab

It's always a good idea to repurpose the stuff you've already got in place, so (logically enough) labs all over the U.S. are being retrofitted to work with nanotechnology. To that end, several Department of Energy national lab facilities are being redesigned as we write to be leading lights for interdisciplinary research on nanotech. In fact, these labs are the basis for a national initiative to promote advances in science, new tools, and enhanced computing capabilities. These new centers will support research at universities and in industry, as well as at other government facilities.

Nanotechnology centers will be showing up at the following existing DOE labs:

- The Center for Nanophase Materials Sciences (CNMS) will be built at Oak Ridge National Laboratory (ORNL) (www.cnms.ornl.gov). The CNMS's three areas of focus will be in nano-dimensioned soft materials, complex nanophase materials systems, and theory/modeling/simulation. The CNMS is geared to take advantage of ORNL's unique capabilities in neutron scattering provided by their very impressive-sounding Spallation Neutron Source and upgraded High Flux Isotope Reactor. This fancy equipment can be used to determine the structure of nanomaterials and to understand synthesis and self-assembly processes in some materials. The CNMS will not be cramped: It's set to be housed in an 80,000 square foot lab and office building with an attached Nanofabrication Research Laboratory.

- The Molecular Foundry will be built at Lawrence Berkeley National Laboratory in Berkeley, California (foundry.lbl.gov). The facility will take a look at the possible sharing of techniques and methods used to fabricate nanoscale-patterned materials. The Foundry will allow scientists from a whole slew of disciplines to perform research in its home in a new laboratory building scheduled to be completed in 2006. The building will contain state-of-the-art instruments and a staff that's savvy in the newest techniques for using those instruments. Folks from academic, government, and industrial laboratories can submit proposals to do their work at The Molecular Foundry.

- The Center for Integrated Nanotechnologies (CINT) is being added to the assets of Sandia National Laboratories (Albuquerque, New Mexico) and Los Alamos National Laboratory (cint.lanl.gov) in 2005/2006. What's unique about CINT, according to their Web site, is their strong emphasis on taking a close look at how you move from scientific discovery to actually using nanostructures in the real world. This involves experimenting with ways to synthesize and process new materials, and how to integrate these materials into structures and test their performance. The plan is for CINT to bring together university faculty, students, laboratory scientists from other national facilities, and corporate researchers to study the integration of new nano-scale materials into unique architectures and microsystems.

Getting Money for Research

Are you ready to apply for a grant to come up with the next great nano-product? Then remember where the federal research money comes from: grants given by individual government departments and agencies in line with their individual missions. Study up on those missions and hone your message accordingly. You'll find more info on the Current Solicitations page of the National Nanotechnology Initiative Web site (`nano.gov/html/funding/currentsol. html`). This particular page offers links to the funding home pages for the departments and agencies involved in the National Nano-technology Initiative (NNI). There's also a solicitation database (`lmisgreen.lanl.gov/ programs`) offered by the Department of Energy Center for Integrated Nanotechnologies. Here you can read all about opportunities for federal funding for nanotechnology research.

But it doesn't stop at grants; there are special programs just to seed commercial activities in nano to stimulate economic growth. These programs push small-business collaboration with universities and other research groups. A page on the National Nanotechnology Initiative Web site (`nano.gov/html/funding/SBIR_ STTRbusinessops.html`) gives you an overview of these federal government programs. The two largest programs are called Small Business Innovation Research (SBIR) and Small Business Technology Transfer (STTR).

If you need help finding your funding, try Tech-Net (`tech-net.sba.gov/`), sponsored by the U.S. Small Business Administration. This Internet-based database includes lists of awards from Small Business Innovation Research (SBIR), Small Business Technology Transfer (STTR), and Advanced Technology Program (ATP). It also includes information on Manufacturing Extension Partners (MEP) centers. If you're looking for a small-business partner, small-business contractors and subcontractors, research partners, or even investment opportunities, check this one out.

✔ The Center for Functional Nanomaterials will be installed at Brookhaven National Laboratory in Upton, New York (`www.cfn.bnl.gov`). The Center is slated to provide cutting-edge resources for research on and fabrication of nano-scale materials. Their emphasis will be on atomic-level customizing of nanomaterials — tweaking specific properties to perform specific functions. The Center should be completed in early 2007 and staffers will get to work by the spring of 2007.

✔ The Center for Nanoscale Materials will be set up at Argonne National Laboratory in Argonne, Illinois (`nano.anl.gov`). This one is a joint effort by the U.S. Department of Energy and the State of Illinois. The CNM will support basic nanotech research and the development of high-end instrumentation. Argonne's Advanced Photon Source (APS) will play an important part. APS's hard X-rays, applied in a nanoprobe beamline, provide a great way to determine the properties of extremely small structures. Argonne has a history of outreach to academic and industrial communities, so they're likely to help support regional and national goals and strategic initiatives. The CNM also welcomes outside users from a wide variety of scientific fields. The hammers should fall silent on this building project around mid-2006 — and operations should begin by the end of that year.

U.S. State and Regional Initiatives

Several states have seen the future, and its name is nano. The amount of resources individual states can devote to bringing us a step closer to this future varies widely, however. For example, one state has fully funded several Nanotechnology Centers of Excellence, while other states have only indicated support of collaboration between universities and corporations within their borders, with little actual funding going on.

One of the big players is the state of New York. Along with some industrial partners, New York committed over $1.4 billion to establish five Centers of Excellence in nanoelectronics, photonics, bioinformatics, information technology, and environmental systems. The combination of the resources of these centers offers a powerful nanotechnology commercialization effort.

As if that weren't enough, New York has also committed over $120 million to build eight Strategically Targeted Academic Research (STAR) Centers and five Advanced Research Centers (ARC). These will not only support research efforts, but also attract some high-powered faculty to New York–based universities. New York is also providing $30 million for activities at 15 Centers for Advanced Technology (CAT) and Regional Technology Development Councils (RTDC) — as well as providing venture capital funding to the tune of $400 million. (If you live in New York, now you know where those tax dollars are going!)

The Massachusetts Nanotechnology Initiative is a pet project of the Massachusetts Technology Collaborative (MTC) (www.mtpc.org). The goal of MTC is to encourage research, new ventures, and new jobs for the Commonwealth. MTC also runs the John Adams Innovation Institute that aims to foster strategic investments that support Massachusetts's knowledge-based economy. For example, the Institute gave a $5 million award to the Center of Excellence in Nanomanufacturing at UMass Lowell. The Center will focus on successful commercialization of nanotechnology, including the challenge of creating the highly trained workforce the discipline demands.

Up in the mountains, the Colorado Nanotech Initiative (www.coloradonano.org) (Colorado Nanotech) is Colorado's nanotechnological arm. Colorado Nanotech fosters opportunities, supports education, increases public awareness, and acts as a general clearinghouse for nanotechnological information and resources around the state and the region.

The Texas Nanotechnology Initiative (www.texasnano.org) naturally thinks big, and so wants to establish Texas as a world leader in the field. They have pulled together a consortium of Texas-based universities, industries, investors, and government representatives to support communication, collaboration, and the sharing of nanotech resources.

California won't be left behind. The Northern California Nanotechnology Initiative (NCNano) (www.ncnano.org) is a regional economic development program. They are working to build what they call a Nanotechnology Cluster in northern California. They have big dreams, saying they'll bring $6 billion in nanotechnology investment and grants into the area, and create 150,000 new local jobs to boot. NCNano hopes to light a nano-fire to inspire the integration of universities, research labs, businesses, and others in the nanotech effort.

The Initiative for Nanotechnology in Virginia (www.InanoVA.org) is another state-sponsored consortium of universities, federal labs, state agencies, and industrial partners. This group promotes research collaboration, workforce growth, technology transfer, and nanotechnology commercialization. Virginia's Center for Innovative Technology (CIT) provided seed money to start the group, along with matching funds put up by partner institutions and corporations.

Minnesota is looking at building a research center at the University of Minnesota to be called OMNI, the Organization for Minnesota Nanotechnology Initiatives (www.nano.umn.edu/omni). OMNI will support research activities in four main categories: nano-scale manipulation and self-assembly; nano-scale characterization and analysis; nano-materials; and nano-devices.

As you might expect, in Pennsylvania, Ben Franklin is leading the charge (Ben Franklin Technology Partners of Southeastern Pennsylvania, that is). This group, along with Drexel University and the University of Pennsylvania, has created NTI, the Nanotechnology Institute (www.nanotechinstitute.org). The NTI's mission is to support "the transfer of discoveries and intellectual knowledge in the area of nanotechnology from universities to industry partners" and to work on "the rapid application and commercialization of this technology to stimulate economic growth."

The linchpin of the New Jersey Nanotechnology Consortium (NJNC at www.njnano.org) is the world-famous Bell Labs nanofabrication laboratory in Murray Hill. This lab offers cutting-edge fabrication capabilities. Joining forces with some top-flight regional academic research institutions and universities (think Princeton, think Rutgers), NJNC has a heady goal: developing the capability to move those exotic nanotechnology ideas from concept to commercialization.

The Connecticut Nanotechnology Initiative (www.ctnano.org) (CNI) is another collaborative effort among industries, universities, and government agencies. Their focus is, naturally, on establishing Connecticut as a leader in nano research, development, and commercialization.

Finally, to prove that too many chefs don't spoil the nano-stew, the Nanotechnology Institute (www.nanotechinstitute.org) is a multistate initiative involving academic and research institutions, corporate partners, private investors, as well as government types interested in supporting economic development. The states participating include New Jersey, Pennsylvania, Maryland, and Delaware — including some international alliances with Japan, Italy, and the United Kingdom thrown in for good measure.

Euro Nano

The trend these days is for Europe to see itself as one big community, whether it involves currency, sports, or technology. As a result, there are a few key Euro-wide organizations driving nanotechnology efforts. Many initiatives seem to still be in the planning stages, but Europe is definitely beginning to get its act together regarding nanotechnology.

The European Commission

The European Union met in March 2000 and called for a better use of European resources and research for technology. They charged the European Commission, an administrative body, with creating a market for science and technology within a disciplinary division called the European Research Area (ERA). A budget program — referred to as a Framework Programme (in this case, FP6) — put aside funds to make this ERA a reality. (FP6 is budgeted to run from 2002to 2006, and it contains a strong focus on nanotechnology.)

If you go to the Community Research & Development Information Service Web site at www.cordis.lu/nanotechnology, you'll get an overview of nanotechnology-related activities across Europe. The site offers information on projects and funding opportunities, as well as information about the European Research Area and the FP6, the sixth Framework Programme.

Goals of the European Commission include research into a range of ambitious nanotech topics:

- understanding phenomena, mastering processes, and developing research tools
- nanobiotechnologies
- engineering techniques
- handling materials and developing control devices
- applications
- development of fundamental knowledge

✔ technologies for production, transformation, and processing

✔ engineering support for materials development

✔ new processes and flexible and intelligent manufacturing systems

The budget allocated for the duration of FP6 is 1.3 billion euros. This budget covers certain technology platforms. The Nanobiotechnologies for Medical Applications, for example, is a technology platform of the ERA that is currently under development with a focus in the medical area. The European Nanoelectronics Initiative Advisory Council (ENIAC— no, it's not the same as that early IBM mainframe) is the European technology platform for nano-electronics, and so on.

To give you an idea of the focus of individual platforms, the stated principal mission of ENIAC (www.cordis.lu/ist/eniac) is to tackle a range of goals:

✔ Provide a strategic research agenda for the nanoelectronics sector, with respect to R&D.

✔ Set out strategies and roadmaps to achieve this vision through the Strategic Research Agenda and other associated documents.

✔ Stimulate increased and more effective and coherent public and private investment in R&D in the nanoelectronics sector.

✔ Contribute to improving convergence between EC, national, regional, and private R&D actions on nanoelectronics within the European Research Area Framework.

✔ Enhance networking and clustering of the R&D capacity in Europe.

✔ Promote European commitment to R&D, ensuring Europe's place as an attractive location for researchers.

✔ Interact with policymakers and other influential people at all levels of society that influence the competitiveness of the sector — such as education and training, competition, finance and investment, and so on.

Keeping folks informed: The Thematic Network

The European Union has also sponsored the Thematic Network (www.cordis.lu/nmp/national-research.htm) to provide a one-stop source of information for all areas of nanotechnology to business, scientific researchers, and communities. A dedicated Web site (www.nanoforum.org) is under development — one that will bring together partners from different disciplines as well as national and regional networks. The idea here is to share best practices in order to stimulate nanotechnology initiatives in less-developed European countries such as Bosnia-Herzegovina and Hungary, encourage young scientists, and spread knowledge and expertise.

EU ministers emphasize the potential of nanosciences and nanotechnologies to spur development in key areas such as healthcare, information technologies, materials sciences, manufacturing, instrumentation, energy, environment, security, and aerospace. These ambitious folks are looking seriously at nano as a significant booster for quality of life, sustainable development, and economic competitiveness.

Bottom line: Europe still seems to be feeling its way in the nano-realm, defining goals, potential programs, and synergies. But keep an eye on the Europeans in the next few years: Their combined energy could make them bigger players in years to come.

Jumping on the Bandwagon: Asia

The United States and Europe are certainly not the only players in nanotechnology. So what's going on in Asian countries? Read on and you'll find out.

Nano in Japan

Nano can make for strange bedfellows, at least from a Western point of view. Japan has given two of its ministries the job of overseeing nanotech research. The Ministry of Education, Culture, Sports, Science and Technology (Sports? Yep.) joins forces with the Ministry of Economy, Trade, and Industry to oversee nanotechnology research. The funding support system is a little complicated, placing two ministers in charge of science and technology.

Japan has a number of different science and technology agendas. They want to promote basic research and development in areas that meet society's needs — the life sciences, for example — as well as information and communications technology, and environmental sciences. Another priority is to support emerging research areas that could have unforeseen benefits.

Development of Japanese nanotechnology as it relates to commercial products probably won't be handled quite the same way as in the United States. Japanese efforts in the nanoelectronic devices area will undoubtedly be controlled by technology giants such as NEC, Hitachi, Fujitsu, and Toshiba, whereas pharmaceutical companies will steer development on the biotechnology side. Japan doesn't currently support nanotechnology with venture-capital dollars, or offer opportunities for small start-ups. Small and medium-size companies may be left behind in the nanotechnology rush, at least in the beginning.

Some key players in Japanese nanotechnology include these:

- ✔ The Council for Scientific and Technological Policy (www8.cao.go.jp/cstp/english/s&tmain-e.html)
- ✔ The National Institute of Advanced Industrial Science and Technology (www.aist.go.jp/index.en.html)
- ✔ The Institute of Physical and Chemical Research (www.riken.go.jp/engn)
- ✔ The National Institute for Materials Science (www.nims.go.jp/eng)

Japanese universities are not being left out of the mix. Tokyo University, the Tokyo Institute of Technology, Osaka University, and Kyoto University, to name a few, are heavily involved in nanotechnology research.

China goes nano

The National Center for Nano Science and Technology (NCNST, as if you really need another acronym at this point) is one of the centers that has received substantial funding from the Chinese government. The National Engineering Research Center for Nanotechnology (NERCN) is another organization favored by Chinese government funding. NERCN borrowed talent from the U.S. in the form of Director Dr. Jie Han, a Chinese-American who was the technical director and manager of NASA's Ames Center for Nanotechnology. Dr. Han also serves as the CEO of the Shanghai National Engineering Research Center for Nanotechnology (SNERC), so this man's plate is very full. The Chinese government has given about 24 million in U.S. dollars to the SNERC.

Current projects that the NERCN and SNERC are working on include:

- ✔ Mass production of nano-diamond coating materials, carbon nanotubes, and nanowires.
- ✔ Nanosensor network systems used for security monitoring.
- ✔ Nanomaterials for environmental protection and healthcare.
- ✔ Chemical and biosensor systems for disease diagnosis.

Other nanotechnology centers in China include the CAS Nanotechnology Research Center, Nanomaterials and Development Application Center, and the Surface Science, Nanotechnology and Engineering Center.

Nano inside India

In 2001, India launched their National Nanoscience and Technology Initiative (NSTI), about the same time as other Asian countries. The program includes three major areas:

- ✔ **Research,** including synthesis and assembly, characterization of properties, and applications.

- ✔ **Education,** supporting advanced schools, symposia, and training workshops for researchers and scholars.

- ✔ **Industry,** emphasizing interaction with industries in a variety of areas, such as nanoelectronics, nanopower/particle production, and surface coatings.

India is building several labs with fancy microscopes and other equipment to analyze and characterize nanomaterials. The main nanotech institutions in India that the NSTI supports are

- ✔ Indian Institute of Science, www.iisc.ernet.in
- ✔ Jawaharlal Nehru Centre for Advanced Scientific Research, www.jncasr.ac.in
- ✔ National Chemical Laboratory, www.ncl-india.org
- ✔ National Physical Laboratory, www.nplindia.org/npl
- ✔ Indian Association of Nuclear Physics, www.iacs.res.in
- ✔ Indiana Institute of Technology, www.iitd.ernet.in

Current research involves applying carbon nanotubes in measuring fluid flow, synthesizing and controlling the properties of nanotubes, using biosynthesis to produce nanomaterials, fine-tuning the biophysics of gene regulation, and looking into manufacturing possibilities such as nanolithography and templates for synthesizing nanowires. (Can't imagine when these guys ever get a break for lunch.)

There has also been a lot of international collaboration going on with India. Their Center for Nanomaterials, for example, was established in partnership with groups in Russia, the Ukraine, Japan, Germany, and the United States.

Nano Is Going Over Big Time in Israel

The Israeli government set up the Israeli National Nanotechnology Initiative (INNI) (www.nanotrust.org.il/inni.asp). Their mission is to make nanotechnology the next big thing in industry by leading the way with a variety of initiatives.

The first step for INNI is to launch partnerships among government, academia, and industry that will allow Israel to step up to the plate in the broader world of nanotechnology. To get there, the INNI has set itself some vigorous goals:

- ✔ Establish a national policy of resource allocation in nanotechnology for optimization of the use of resources and faster realization of market viability.

- ✔ Formulate a long-range nanotechnology program for research and technology development in academia and industry, and promote the establishment in Israel of a world-class infrastructure to support it.

- ✔ Actively seek resources from public as well as from private sources to fund research, technology projects, and infrastructure projects.

- ✔ Lead the selection of the projects per agreed national priorities, allocate the budgets, and review progress on approved projects.

- ✔ Promote the establishment of local nanotechnology-based industry, which will make an impact on the economic growth of Israel and benefit the investors.

The Israel Nanotechnology Trust (www.nanotrust.org.il), known as the INT, is the part of the Israeli National Nanotechnology Initiative (INNI) that raises and distributes money.

Another group active in Israel is the Consortium for Nano Functional Materials (NFM). This organization consists of participants from 14 industries and 12 academic research groups. NFM members are performing basic R&D in the field of nano-materials and technologies. The NFM seeks to expand their activities by working with groups from around the world.

The INNI has stated that to gain a leadership role, Israel has to invest at least $300 million in research and development over five years. They've made a good start: Their current total investment is around $150 million.

To search out this kind of funding, the Israeli Nanotechnology Trust is looking at a variety of sources, including Jewish organizations. Traditionally, money from Jewish groups goes toward supporting immigration to Israel, helping to develop agriculture, and building up military strength. The message they present to Jewish donors is that an emphasis on funding science and technology — especially nanotechnology — is the way to jumpstart Israel's future.

Can a smaller country such as Israel hope to compete in the nanotechnology arena? Probably not. Smaller countries don't have a prayer of mustering the spending power of the United States, the European Union, and Japan. Still, they may have a shot at becoming leaders in applications of nano that focus on solving their particular problems, such as water and energy shortages.

No wonder Israel is applying nanotechnology to areas of clear practical importance to national survival. For example, water desalination, energy, biotechnology, and semiconductors all have a strong social and/or economic focus in Israel.

Bottom line: As the components of technology get smaller and smaller, world-wide interest in developing them gets bigger and bigger.

Chapter 14

Nanotechnology Goes to School

. . . .

In This Chapter

▶ Looking at schools that are into nanotechnology in a big way

▶ Understanding some of the ties of industry and government to academia

▶ Exploring educational opportunities in nanotechnology

▶ Discovering a handful of other academic hotbeds of nano in the U.S. and elsewhere

. . . .

*N*anotechnology is still largely in a research phase — and a great deal of that is being done on university campuses. Because a lot of this research work requires some big bucks — not in big supply on many campuses today — much of this work goes on in conjunction with business and governments.

In September of 2001 (for example), the U.S. government selected six schools as Centers of Excellence in Nanotechnology. (These are included in the upcoming list, along with a few others.) The National Science Foundation has several partnerships with academia (several of which we mention here). But more than a hundred schools now have nanotechnology research underway — and that's just in the United States. Japan, China, and Israel (to name a few) are also active.

In this chapter, we pinpoint several universities going small in a big way.

Harvard . . . of course

One of the designated Centers of Excellence in Nanotechnology, Harvard is no slacker when it comes to all things small.

Harvard's Materials Research Science and Engineering Center (www.mrsec. harvard.edu) brings together 21 faculty members from the departments of engineering, physics, biology, chemistry, and even Harvard Medical School.

The three main research groups are in hot pursuit of distinct projects:

- ✔ **Multiscale mechanics of films and interfaces.** In plain English, these folks are looking at the behavior of thin films.

- ✔ **Engineering materials and techniques for biological studies at cellular scales.** This group is studying the structure and properties of the materials that compose biological cells.

- ✔ **Interface-mediated assembly of soft materials.** This research involves controlling various aspects of self-assembly processes.

The Materials Research Science and Engineering Center sponsors an industrial internship that allows students involved in nanotechnology research to work hand in hand with industry researchers.

Individual departments at Harvard are also immersed in nanotech research in areas such as computer devices, chemistry, nano-size switches and wires, and molecular imaging. Harvard Business School, not to be left behind, even explored investment opportunities for nanotechnology at their Annual Principal Investment Conference in 2005.

Small as Rice

Rice University in Texas is home to (among others) Richard Smalley, Nobel Laureate in Chemistry with an impressive reputation in nanotechnology research. Smalley heads up The Smalley Group at Rice.

The Center for Nanoscale Science and Technology at Rice (cnst.rice.edu/ whatshot.cfm) involves 14 academic departments, ranging from Earth Science to the Jones School of Management to Chemistry and even to Religious Studies.

The Center focuses on three areas: *wet, dry,* and *computational* nanotechnology. Here's how these break down:

- ✔ **Wet nanotechnology** is the study of biological systems that live in the water. This is the area where genetics, enzymes, and various cellular components are studied.

- ✔ **Dry nanotechnology** relates to surface chemistry and materials fabrication, including the study of nanotubes, silicon, and other inorganic materials. Dry nanotechnology can involve metals and semiconductors (which its "wetter" cousin can't).

- ✔ A **computational** focus involves modeling and simulation of very complicated nanometer-scale structures. The study of nano-computation can have a huge impact on the evolution of both its wet and dry brethren.

Rice often works with Texas-based NASA on projects involving SWNT (single-walled nanotubes) and sci-fi-sounding stuff like pulsed-laser vaporization (a process used to produce SWNTs).

Small Things in the Big Apple: Columbia

The Center for Nanostructured Materials at Columbia University (research.radlab.columbia.edu/mrsec) was established by a National Science Foundation grant way back in 1998. Their work involves teams of academics, people from industry, and laboratory scientists developing complex metal-oxide nanocrystals into thin films that can be used to improve the strength of materials. The center is also looking into structurally integrated chemical films that contain nanoparticles.

There's a lot of educational outreach from this school, including a summer program for undergraduates who want to explore nanotechnology research. They also sponsor a program for K-through-12 science teachers in the New York City area. Finally, the folks at Columbia have created a resource-sharing program called Instrumentation User Facilities to make certain difficult-to-find (and hard-to-afford) equipment available to others for nanotechnology research.

Columbia also received an initial U.S. government grant of $10.8 million to fund the work of 16 researchers for five years in their Center for Electronic Transport in Molecular Nanostructures. Professors Ronald Breslow (who won a National Medal of Science) and Horst Stormer (a Nobel Prize winner in physics) lead the center.

Visit the nanoportal at Columbia at www.cise.columbia.edu/nano/ to find out all that's going on there.

"And Perhaps Cornell?"

Cornell was named as 1 of 13 members of the National Nanotechnology Infrastructure Network (www.nnin.org/), a national nanotechnology laboratory network, in December 2003. This consortium was funded to the tune of about $70 million over five years, beginning in January of 2004. Figure 14-1 provides a map that shows the locations of all 13 member schools.

Folks at Cornell focus on materials science and engineering research, including nanostructured materials in electronic and mechanical devices.

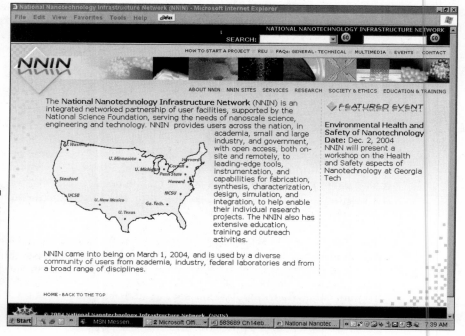

Figure 14-1:
The
National
Nano-
technology
Infra-
structure
Web site.

The Applied Physics faculty at Cornell are busy researching a variety of nano-related topics, including

✔ The dynamic mechanical properties of nano-scale materials

✔ Growth mechanisms of nanomaterials

✔ Properties of nanocomposites

✔ Nano-scale electronic and magnetic phenomena

✔ Development of nano-scale devices

✔ Thin films

✔ Development of applications for light-emitting nanoparticles called *CU dots* — not to be confused with quantum dots (which we discuss in Chapter 8), each of these contains a core of fluorescent dyes inside a glass shell

✔ Self-assembly at the molecular level

Under the direction of Professor Bruce Lewenstein, Cornell's Department of Communication has even teamed up with their Science & Technology Studies department to investigate the social and ethical issues posed by nanotechnology. (Visit www.people.cornell.edu/pages/bvl1/NanoSEI.htm to read more about this effort.)

Check out the Ask a Scientist feature at the Cornell Center for Materials Research Web site (www.ccmr.cornell.edu/education/ask/). Not all the questions asked and answered here concern nanotechnology, but if you have a question about a nano-related topic, you can post it here and get an answer from a real live scientist.

Nano House on the Prairie: Northwestern University

Northwestern's Institute for Nanotechnology (www.nanotechnology.north western.edu/) is a kind of umbrella group for administering multimillions in nanotechnology research dollars.

One of the investments recently at Northwestern was to build a 40,000-square-foot Center for Nanofabrication and Molecular Self-Assembly on its Evanston campus. The U.S. Department of Health and Human Services funded it (to the tune of $14 million), which makes it one of the first federally funded facilities dedicated to nano.

Other initiatives at Northwestern include these:

✔ Nanoscale Science & Engineering Center for Integrated Nanopatterning and Detection Technologies. This is one of seven Nanoscale Science and Engineering Centers (NSEC) funded by the National Science foundation at various universities. You can find a listing of all the NSECs at www. nano.gov/html/centers/nnicenters.html.

✔ Nanoscale Interdisciplinary Research Team for Multiscale Modeling and Catalysts for Green Chemistry and Engineering (try fitting *that* on your educational resume).

✔ NIAID Specific Detection of HIV Targets by Gold Nanoparticle Probes.

✔ Center for Transportation Nanotechnology.

Northwestern faculty have laurels but certainly aren't sitting on them. John Pople won the Nobel Prize in Chemistry in 1998, Chad Mirkin pioneered dip pen lithography (a new method of "writing" at the molecular-scale), and Richard Van Duyne is an Alfred P. Sloan fellow in Chemistry who has won numerous awards in the area of the dynamics of surface processes.

Small Progress at Rennsselaer

In Troy, New York, plucky little RPI (Rennsselaer Polytechnic Institute) has established a nanotechnology center that is hard at work on research and education initiatives. Here they study advanced coatings for materials, the nano dimension of biosciences and biotechnology, and nanoelectronics, as well as nano-scale mechanics and systems.

RPI hosts the Nanoscale Science and Engineering Center for Directed Assembly of Nanostructures (www.rpi.edu/dept/nsec/about_us.html) along with the University of Illinois and Los Alamos National Laboratory. (The Center's Web site is pictured in Figure 14-2.) Established in 2001 by the National Science Foundation, the effort is housed in the Rennsselaer Nanotechnology Center. The Center is focusing on two areas; the assembly of nano-sized materials into gels and polymer nanocomposites, and nanostructured biomolecule composite architectures. (Nanocomposites might be used in, for example, stain-resistant fabrics or to create stronger plastic bags.)

The RPI-Industry Partnership in Nanotechnology provides support for industry-related research with companies such as Eastman Kodak (in Rochester, New York) and IBM (in Poughkeepsie, New York), as well as more distant partners like Philip Morris (in Richmond, Virginia).

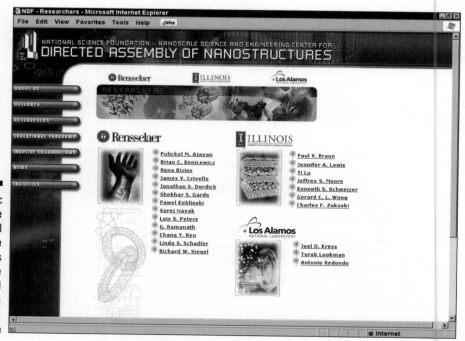

Figure 14-2:
The National Science Foundation's Nanoscale Science and Engineering Center site.

Ben Gurion University and Nano

Israel is a hotbed of nanotechnology, where researchers have developed the ability to produce large quantities of metal nanopowders for use in industry (far exceeding the rate at which the U.S. can produce them). Ben Gurion University is one group leading the way in Israeli academia.

Search the Ben Gurion University Web site (www.bgu.ac.il/) and you get hundreds of listings of courses and articles related to nanotechnology. This school has a $100 million center for nanotechnology, so you know they have definitely caught the religion of small.

The people at Ben Gurion are working on everything from carbon nanotubes to thin films to DNA manipulation to polymer matrices to quantum dots. They are a part of the Nano-Functional Materials Consortium, comprising 14 companies and 12 academic research organizations in Israel and working on the use of nanotechnology in manufacturing processes.

Check out the Ilse Katz Center for Meso and Nanoscale Science and Technology at www.bgu.ac.il/nanocenter/index.html for a list of labs run by specific professors as well as a rundown of various activities and events related to nanotechnology.

Technion, The Israel Institute of Technology (www.technion.ac.il) is at once the oldest institute of technology in Israel, started in the 1920s, and one of the most cutting edge. They are very active in the nanotechnology research area and will open the Sara and Moshe Zisapel Nanoelectronic Center (www.ee.technion.ac.il/Labs/Nano/) in 2005.

Made in Japan: University of Tokyo

Japan is where nanotubes were first discovered, so don't be surprised if other advances come from this small — but technologically advanced — country. The Quantum-Phase Electronics Center at the School of Engineering, University of Tokyo (www.qpec.t.u-tokyo.ac.jp/) is looking into the electronic properties of SWNTs (single-walled nanotubes) in a big way. They were founded in 2001 to study the relationship of quantum physics to engineering.

The center takes its name from its study of materials — such as nanotubes — in which electrons are in a "quantum-phase" (that is, are ruled by the laws of quantum mechanics). The center seeks to combine the necessary resources in science and engineering to develop a new electronic based on such materials and devices. Their goals include ultrafast optical switching, high-temperature superconductors, and the controlled use of electron orbitals (required for molecular electronics).

California (Nano) Dreaming at Berkeley

With its traditional tie-ins with some pretty famous research labs such as Lawrence Livermore National Laboratory, Berkeley is a natural front-and-center location for such a research-intensive field such as nanotechnology. The Berkeley Nanotechnology Forum, held in April 2005, is just one effort of this group to lead the way into the nano-realm.

The Nanotechnology Initiative Research Group at Berkeley (www.me.berkeley.edu/nti/home.html) is a part of the Department of Mechanical Engineering that deals with nano-scale engineering, nanocomposites, and nanowires, among other weighty topics.

As an example of the type of people who pursue nanotechnology at Berkeley, the Chair of the department, Professor Albert Pisano, also happens to be head of The Nanotechnology Steering Committee of the American Society of Mechanical Engineers. He's a big fish in the nanotech pond, having served as the Program Manager for MEMS for the Defense Advanced Research Projects Agency (DARPA). A researcher in the chemistry department, Paul Alivisatos, was the founding editor of *Nano Letters,* a scientific journal devoted to nanotech research. (He's also scientific founder of Nanosys Inc., a startup specializing in nanomaterials.) Alivisatos is currently the director of Berkeley Lab's Molecular Foundry, one of five Department of Energy research centers.

Other departments at Berkeley active in nanotechnology include the Physics Department — busily studying vibration requirements for environments where nano devices are being manufactured — and Electronic Engineering — bent on exploring nanotechnology uses in the semiconductor industry.

Visit nano.berkeley.edu for more about what's going on at Berkeley, including their leadership role in COINS (The Center of Integrated Nanomechanical Systems).

Educating Yourself in Nano

If your goal is to study nanotechnology, there are all kinds of resources out there. Start with the National Science Foundation's National Nanotechnology Initiative's Web site (nano.gov). Here you can use the Education Center link (on the left side of the home page) to get information on nanotechnology facts for grades K–12 (see Figure 14-3) or get a listing for universities that grant degrees in nanotechnology. You might also check out the Institute for Nanotechnology at Northwestern University (www.nsec.northwestern.edu/education.htm), whose goal is to "foster a lifelong interest in science and technology" by teaching people all about nanotechnology.

Women in Nanotechnology

Though many of the people teaching and researching in nanotechnology today are men, women are also active in the field. A case in point is Angela Belcher, an MIT materials scientist and winner of a 2004 MacArthur Fellowship (sometimes referred to as a "genius grant"). In addition to her work at MIT, she has founded a company called Cambrios Technology, which does work in microchip fabrication and other areas. Concerned with spreading nanotech knowledge, she has worked with local schools to make her laboratory at MIT available to help educate others. If you or somebody you know is interested in studying nanotechnology, rest assured there is opportunity there for everyone!

At the time of this writing (according to the NSF National Nanotechnology Initiative), the following U.S. universities grant degrees in nanotechnology:

- In conjunction with the University of Pennsylvania, you can get an Associate Degree in nano-biotechnology at community colleges in Pennsylvania.

   www.nanotechinstitute.org/nti/workforceDevelopment.jsp

- Dakota County Technical College (Rosemont, Minnesota), in conjunction with the University of Minnesota, offers an Associate in Applied Science degree in Nanoscience Technology.

   www.dctc.mnscu.edu/programs/nano.htm

- Louisiana Tech University offers a Master of Science degree in Molecular Sciences and Nanotechnology.

   www.coes.latech.edu/grad/msnt/index.htm

- Rice University offers a Professional Master of Science in Nanoscale Physics.

   www.profms.rice.edu/nanoscalePhysics.cfm

- University at Albany-SUNY's College of Nanosciences and Nanoengineering offers a Ph.D. and M.S.

   www.albany.edu/grad/school_nanosciences_nanoengineering.html

- University of Washington grants a Ph.D. in Nanotechnology.

   www.nano.washington.edu/about/index.html

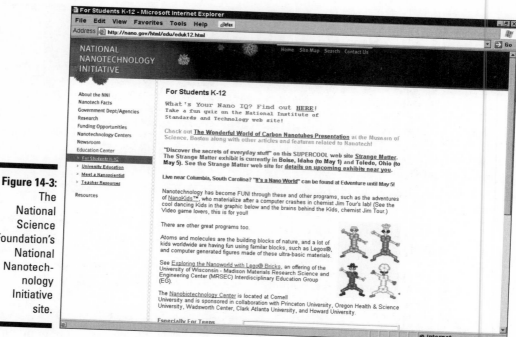

Figure 14-3:
The
National
Science
Foundation's
National
Nanotech-
nology
Initiative
site.

And a Whole Bunch More . . .

These days, so many universities are active in nanotechnology, it's impossible to whittle them down to ten. Here are some schools to keep an eye on in the United States:

- **University of Pennsylvania at Philadelphia, Pennsylvania:**
  - Penn Regional Nanotechnology Facility:

    www.seas.upenn.edu/nanotechfacility/
  - Nano/Bio Interface Center:

    www.nanotech.upenn.edu/
  - Info page on research at UPenn in Nanotechnology Research News:

    www.upenn.edu/ben-penn/nanotech.html
- **Massachusetts Institute of Technology at Boston, Massachusetts:**
  - Nanostructures Laboratory:

    nanoweb.mit.edu
  - Institute for Soldier Nanotechnologies:

    web.mit.edu/isn

Space Nanotechnology Laboratory:

`snl.mit.edu`

MIT News Office page on Nanotechnology:

`web.mit.edu/newsoffice/topic/nanotech.html`

✔ **Stanford University at Stanford, California:**

- Nanofabrication Facility:

  `http://soe.stanford.edu/research/lab_ctr_dtl.php?org=30`

- Nano-Photonics Laboratory:

  `http://soe.stanford.edu/research/lab_ctr_dtl.php?org=91`

- Nanocharacterization Laboratory:

  `http://soe.stanford.edu/research/lab_ctr_dtl.php?org=196`

✔ **University of California at Santa Barbara, California:**

UCSB Nanofabrication Facility: `www.nanotech.ucsb.edu/`

Center for Biologically Inspired Nanocomposite Materials

✔ **Arizona State University at Tempe, Arizona:**

General information: `www.asu.edu/`

Nanostructures Research Group: `www.eas.asu.edu/~nano/`

In other parts of the world, look for nanotech progress from these schools:

✔ **Tokyo Institute of Technology, Japan:** `www.titech.ac.jp/`

✔ **Osaka University, Japan:** `www.osaka-u.ac.jp/`

✔ **The Technical University of Delft, Netherlands:** `www.tudelft.nl`

✔ **University of Kent, England:** `www.kent.ac.uk/`

✔ **University of Wurzburg, Germany:** `www.uni-wuerzburg.de/?lang=en`

✔ **University of Copenhagen, Denmark:** `www.ku.dk/`

✔ **University of Antwerp, Belgium:** `www.ua.ac.be/main.asp?c=*ENG`

✔ **University of Science and Technology of China:** `www.ustc.edu.cn/en/`

Part VI
The Part of Tens

The 5th Wave   By Rich Tennant

"...and I think we all owe a big thanks to Doug Gretzel for all his work in helping to develop our nano-robotic Pimple Popper."

In this part...

This is the spot in a *For Dummies* book where we get into listmania. This part offers two chapters that give you top-ten lists of many things nano.

Chapter 15 looks at some of the people behind the hype — the movers and shakers in the world of nanotechnology — from the brainy scientists making discoveries to the businesspeople leading the way in nanotech investing. Chapter 16 gives you ten-plus resources — including Web sites and magazines — to use when you want to delve farther into nanotechnology.

Chapter 15

Ten (or So) Nanotech Movers and Shakers

> *Nanotechnology is the art and science of building stuff that does stuff at the nanometer scale.*
>
> — Richard Smalley

We couldn't have said it any better ourselves.

Nanotechnology has developed its fair share of notable overachievers — folks who are creatively exploring (and staking claim to) the uncharted territory of the nano scale. In this chapter, we're going to highlight a few such notables. This list of movers and shakers comprises scientists and business leaders — and even scientists who *are* business leaders. Although this list exceeds its quota (sixteen!), it's certainly not complete; it continues to evolve.

And the winners are . . . in no particular order . . .

Richard Smalley

Richard Smalley is a professor at Rice University, situated in the Great State of Texas (Houston, Texas, to be more precise). In 1985, Smalley, along with Robert Curl and Sir Harold Kroto, discovered C_{60}, the buckminsterfullerene (also known affectionately as the buckyball) — in effect, the key to the molecular structures most readily used in nanotechnology. As a result, all three shared the Nobel Prize in 1996 in chemistry.

Professor Smalley has spent the last 20 years or so busily promoting nanotechnology in general, as well as perfecting production methods of another form of fullerene, the carbon nanotube — "a gift from Mother Nature," as he describes it. In 2000, he founded Carbon Nanotechnologies Inc. (CNI), a start-up company devoted to large-scale production of single-walled carbon nanotubes. A staunch advocate of nanotechnology, Smalley testified before congress in 1999 in support of what later became the National Nanotechnology Initiative (NNI). In addition, he has promoted an energy overhaul on a scale as big as the Apollo space program, utilizing carbon nanotube fibers for efficient energy distribution. He encourages young students to seek careers in the sciences with his straight-ahead mission statement: "Be a Scientist, Save the World."

Charles Lieber

A Harvard University professor, Charles Lieber has pioneered the synthesis, characterization, and development of nano-scale wires. He has continued to demonstrate applications of these nanowires in nanocomputing, nanoelectronics, nanophotonics, and biological and chemical sensing. Capitalizing on these breakthroughs in nanomaterials, he founded a nanotechnology company — Nanosys, Inc. — in 2001. Its goal was to develop a "google" of patents (over 350 completed patents and patent-pending applications). These would include flexible electronics (think circuit boards on thin sheets of plastic), solar cells, fuel cells, and nonvolatile computer memory. Not only is Lieber making nanotech materials and applications, he has also developed a new, chemically sensitive microscope for probing organic and biological material at the nano scale.

Hongjie Dai

Stanford University associate professor Hongjie Dai has taken carbon nanotubes to new heights — or (perhaps more appropriately) depths — of nanosophistication. After his stint at Rice University, where he worked with Richard Smalley, he has continued to study the suitability of carbon nanotubes for future miniaturized devices. No easy task. For openers, he not only has to achieve precision operations despite the small size of the carbon nanotubes, but also decipher their unique quantum effects. As it happens, Dai's team is making progress with two applications that will improve nanotech itself. They're using nanotubes as chemical sensors and as tips for atomic-force microscopes — both of which give nanoscientists unprecedented resolution and sensitivity to see what they're doing down there.

James Heath

California Institute of Technology professor James Heath started his nanotechnology trek back in 1985; as a graduate student, he ran the experimental apparatus in the work that led to Richard Smalley's discovery of C_{60} at Rice University. He has the happy knack of getting his experiments to work — and is described as a brilliant experimental scientist by his peers. After a brief stint at IBM's T. J. Watson Research Labs, he moved on to the University of California at Los Angeles, where he pioneered the molecular switch, using nanowires and molecules. Additionally, he developed a scanning optical microscope used in noninvasive probing of the electrical functions of living cells — in effect, poking and prodding cells to image them without hurting them. Now, at Caltech, he is combining his knowledge of molecules, nanowires, and biology to come up with a single cell-sensing device (that is, a lab-on-a-chip) in conjunction with Stephen Quake, a Stanford University microfluidics expert.

His hard work has not gone unnoticed. Among many awards, he's won the Feynman Price in Nanotechnology in 2000, and was even seen alongside Bruce Springsteen in *Vanity Fair* magazine's Hall of Fame in 1999 — maybe that means nanotech was born to run.

James Von Ehr II

James Von Ehr is the founder, chairman, and CEO of Zyvex Corporation, a nanotech company specializing in nano-size manipulators — tools that allow scientists to manipulate nano-size structures under a microscope. Before founding Zyvex, Von Ehr was founder, president, chairman, and CEO of Altsys Corporation, the company that developed the first PostScript drawing program. Altsys was sold in 1995 to Macromedia (the company that unleashed Flash animation on all us unsuspecting visitors to Web sites across the Internet). Clearly this shows that Von Ehr has a knack for commercializing technology — but more importantly this (major) sale brought in enough major *bucks* to start a nanotech company — which is inevitably a huge investment.

Along with running Zyvex since 1997, he is a founder of the Texas Nanotechnology Initiative and is on the Board of Directors for the NanoBusiness Alliance. In 2003, Von Ehr testified before the Senate to promote the 21st Century Nanotechnology Research and Development Act, which was signed into law later that year.

George Whitesides

George Whitesides is a chemistry professor at Harvard and a member of the Nanotechnology Technical Advisory Group — a U.S. government advisory committee. His research has influenced (and continues to influence) material science, surface science, microfluidics, self-assembly, and of course, nanotechnology. His main focus has been surface chemistry — particularly that of organic surfaces — nonmetal surfaces such as skin, wood, or fabric. Organic surfaces are highly disordered, making their interfaces with other surfaces difficult to study because collecting information is tricky. Whitesides pioneered an unusual approach that used self-assembled monolayers (SAMs) to show ordered organic surfaces. This made it possible to *design* surfaces at the molecular scale — and (an added bonus here) it's a very flexible technology. Lots of variations are possible. Another claim to fame for Whitesides is his development of *soft lithography,* a process for molding and printing micro- and nano-structures. This technique makes possible smaller electronics and microfluidics (lab-on-a-chip components).

Paul Alivisatos

Paul Alivisatos is a chemistry professor at the University of California, Berkeley, as well as a researcher at Lawrence Berkeley National Laboratory. His real claim to fame lies in his work with semiconducting nanocrystals. These nanocrystals come in different shapes and sizes — quantum dots, nanorods, tetrapods, and other such exotic creations. The quantum-dot connection led him to help launch Quantum Dot Corp. in 1998 with their first product, Qdots (available in 2002 for use in bio-labeling). He incorporated his nanorods for use in plastic-nanocrystal hybrid solar cells. And there hangs a tale. It turns out that since the nanorods were randomly mixed for mass production, they weren't especially energy efficient. That led to his next discovery — *tetrapods* — pyramid-shaped nanocrystals that resemble children's jacks. When these crystals align, each one has at least one peg that stands up — which increases their efficiency in converting sunlight to usable energy.

With such raw talent, nothing could hold him back. Other entries soon made their way onto his résumé, including these:

- Founding editor of Nano Letters, a scientific journal devoted to nanotech research
- Scientific founder of Nanosys Inc., a start-up specializing in nanomaterials
- Director of Berkeley Lab's Molecular Foundry, one of five Department of Energy research centers

Angela Belcher

Biomedical engineer Angela Belcher is an Associate Professor of Materials Science at MIT. Before coming to MIT, she was a chemistry professor at the University of Texas in Austin, where she pioneered the (cost-effective) use of genetically modified viruses in the self-assembly of nanowires, thin films, and other nanomaterials. In essence, she's using Mother Nature to help build nanostructures. Belcher has successfully combined inorganic and biochemistry, molecular biology, electrical engineering, and material science into one lean, mean nanotechnological machine.

Her work promises a direct (and sizable) impact on drug discovery and delivery, materials and catalysts, and self-assembling electronic materials. For example, she has coaxed viruses to crystallize and retain their structure indefinitely — providing a new approach for preserving vaccines without refrigeration. This approach may well revolutionize the way vaccines are distributed to other parts of the world, particularly to developing countries where refrigeration may be in short supply. (The added bonus here is that these viruses are harmless to humans and animals.) Other applications include making thin films possible for the stable storage of proteins and DNA. (Think drug delivery.)

Professor Belcher has won numerous awards and accolades, including the prestigious 2004 MacArthur Foundation award — also known as a "genius grant." She is also co-founder of the nanotech start-up Cambrios Technologies Corp., which uses biomolecules to produce better, cheaper electronic devices.

Visionaries: Richard Feynman and Eric Drexler

Once described as "The Smartest Man in the World," Richard Feynman laid out the essentials of a nanotech capability in his 1959 talk, "There's Plenty of Room at the Bottom." Educated at MIT and Princeton, he started his career as a group leader for the Manhattan Project in his early 20s. In 1950, he moved on to Caltech. His highly effective teaching broke down problems and concepts to their simplest level, and guided students to discover the answers themselves. (Hmm . . . maybe he should've written this book? Oh well. Glad he left some of the fun for us.) He won the Nobel Prize in Physics in 1965, sharing it with Julian Schwinger and Shin'ichiro Tomonaga, for work on how subatomic particles interact. In 1986, he worked on the commission investigating the *Challenger* space shuttle explosion, accurately and simply demonstrating the cause of the disaster. Although he died in 1988, his vision for nanotechnology's potential lives on.

Also from MIT, Eric Drexler illustrates molecular manufacturing and lays the groundwork for the public's current perception of nanotechnology some of which is still, um, mired in speculation) in his 1986 book *Engines of Creation: The Coming Era of Nanotechnology*. This is the book that first mentions "gray goo," warning of self-replicating nanotechnology running amok and covering the earth. Science fiction, always fond of farfetched disasters, grabbed the topic and ran with it. Most scientists downplayed this scenario, but Drexler's book sparked a wider interest in nanotechnology and brought it to the attention of the public. Also in 1986, Drexler founded the Foresight Institute to prepare society for emerging technological revolutions. He is currently a Research Fellow at the Institute for Molecular Manufacturing, a sister organization to the Foresight Institute, promoting the science and engineering of molecular manufacturing.

Nanoshells: Naomi Halas and Jennifer West

Drs. Halas and West have continued to make progress with nanoshells that can find and "cook" cancer cells. As we describe in Chapter 11, gold-coated silica spheres enter the bloodstream and attach themselves to cancer cells — and when illuminated with a laser, the nanoshells give off enough heat to kill the tumor cells. Naomi Halas invented nanoshells in the 1990s; along with Jennifer West, she applied them to cancer therapy. Since then, they've continued to test this capability, seeking FDA approval for human use. They both co-founded Nanospectra Biosciences as a way to market their research. Their work is considered the "Best Discovery of 2003" by Nanotechnology Now Web site and they were finalists for *Small Times* magazine's 2004 Nanotechnology Researcher of the Year. Separately, Jennifer West's research in synthesizing blood vessels won *Technology Review* magazine's 2003 TR100 Top Young Innovators award.

Molecular Logic: James Tour and Mark Reed

James Tour is a chemist at Rice University. Mark Reed is a physicist at Yale University. Together, they make molecular logic. While Tour was at the University of South Carolina, he collaborated with Reed throughout the 1990s. Tour would synthesize the molecules and Reed would perform the experiments. They coax the molecules into spontaneously orienting themselves onto the electrodes. If this approach truly works out, it may be a cheap replacement for silicon-based computer chips. Additionally, since they incorporate molecules, these innovations could bridge the gap between biology and computing — resulting in implantable biochips that respond to chemical clues and discharge an appropriate dose of medication.

Regardless of long-term applications, Tour and Reed were thinking big and small at the same time when they co-founded Molecular Electronics Corp. (MEC) in 1999. Their goal was to develop a viable complement to silicon — molecules working alongside silicon, the material used in our computers. Since 2002, MEC has identified over 15,000 circuit connection devices performing a variety of physical, chemical, and production tests and analyses.

Investors: Steve Jurvetson and Josh Wolfe

Steve Jurvetson is the managing director at Draper Fisher Jurvetson, a venture capital firm — and got there on a fast track. Jurvetson whizzed through Stanford undergrad as an electrical engineer (first in class), stayed on for his master's degree, and finished it all off with another degree from Stanford's Business School. He matches this scholarly knowledge with calculated risk and long-range vision. Not only did he make partner rank with Draper Fisher within six months of arriving (in 1996), he backed Hotmail when it was a start-up — and sold it to Microsoft, two years later, for $400 million. Voted *Fortune* magazine's "Brain Trust of Top Ten Minds," Jurvetson is a regular proponent of nanotech as co-chairman of the NanoBusiness Alliance.

Josh Wolfe is co-founder and managing partner of Lux Capital, a venture capital firm focusing on nanotechnology investments. He has a long and established financial history with Salomon Smith Barney, Merrill Lynch, and Prudential Securities. He is a vocal proponent of nanotech, speaking as a guest at Harvard, getting himself quoted in *The Wall Street Journal.* Wolfe is also a co-founder and advisor to The NanoBusiness Alliance and a Senior Associate of the Foresight Institute for Nanotechnology (Eric Drexler's organization). He has been honored by *Crain's Magazine* as one of "The 40 under 40" in 2003 — and has made a splash as author of the "Nanotech Report" and Editor of "Forbes/Wolfe Nanotech Report."

Chapter 16

Further Reading on the Web and in Your Library

Nanotechnologyfordummies.com . . . is your go-to source for information on nanotechnology and this book!

— Richard Booker, author of *Nanotechnology For Dummies*

Nothing like a little shameless self-promotion to start off a chapter. And so begins your journey of self-discovery into the world of nanotechnology, where you put down this book (reluctantly, of course) and log on to the Web and/or pick up a magazine. At this point in its history, nanotechnology is quickly evolving — the changing research and breakthroughs are coming fast and thick. We give you a solid foundation in this book, but it's up to you to keep up with this "Next Industrial Revolution" as it unfolds. Good luck!

Web Sites

Nanotechnology is easy to look up on the Internet — just type **nanotechnology** using your favorite search engine. Better yet, take a look at these Web sites first and save yourself some time and trouble.

www.nanotechnologyfordummies.com

Hey! How 'bout that! This book has its own Web site . . . although the name *is* a bit long. Regardless, this is your go-to source for information on nanotechnology and this book! This Web site will field questions and provide updates on nanotechnology — with the ultimate goal of preparing the way for a second edition of this book.

nanobot.blogspot.com

Howard Lovy's Nanobot blog is a good source for up-to-date information on things nanotechnological. Howard scours the Internet, looking for anything nanotech-related, and invites viewers to expand upon his more-than-able synopses by posting comments. This site provides surprisingly good compilations and coverage. Consider it your second "go-to source" for nanotech information.

www.azonano.com

Touted as "The A to Z of Nanotechnology," this Web site provides a wealth of knowledge about anything nano. Answers to virtually any question you may have on nanomaterials, applications, or industries are laid out from A to Z. The site is very easy to navigate, useful, and (best of all) free.

www.nano.gov

This is the United States Government's National Nanotechnology Initiative Web site. It details the initiative's background, lists the government agencies involved with nano, and points you to research and university centers doing nano stuff. (There's even a Teacher Resources section.) It is certainly a good starting point for determining the government's flow of money and which research is being pursued.

The Sandia National Laboratory maintains another U.S. Government Web site devoted to nanoscience. (Check it out at `nano.sandia.gov`). It has some very good descriptions — and, if you have high bandwidth, good videos. Check out, for example, the videos of micro-electromechanical systems (MEMS) as they interact with dust mites (`www.mems.sandia.gov`).

www.forbesnanotech.com

The Forbes/Wolfe Nanotech Report's Web site provides some great, easy to read, semi-technical articles expanding upon topics you may have read about in this book. Some portions of the Web site are subscriber-only (with a hefty price tag), but they also have a lot of good stuff that's free. The "Wolfe" portion of Forbes/Wolfe is Josh Wolfe, one of the Movers and Shakers we mention in Chapter 15. Two other nanotech business Web sites are

✔ **www.nanobusiness.org:** The NanoBusiness Alliance is the first industry association promoting nanotechnology. This Web site gives you a heads up on some of the business applications that are being developed, along with conferences you may attend.

✔ **www.phantomsnet.com:** Cientifica, the brains behind www.phantomsnet. com, provides in-depth nanotechnology reports and consulting expertise targeting large companies. Although the reports are very expensive, the site also provides some free news articles — a great resource for Fortune 500 CEOs looking to expand their business into nanotech.

www.fda.gov/nanotechnology

If you're into nano-health, the United States Food and Drug Administration (FDA) has a section devoted to nanotechnology. Their Web site outlines how it plans to ensure that nanotechnological products, whether they're in the form of cosmetics, drugs, food, or medical devices, are both safe and effective. The National Cancer Institute (nano.cancer.gov) has a Web site that lays out the role of nanotechnology in the diagnosis, prevention, and treatment of cancer.

www.nano.org.uk

The Institute of Nanotechnology, based in the United Kingdom, provides worldwide coverage of technological developments working closely with governments, universities, and companies. For the CEO, they have nanotechnology reports on the U.K., Europe, North America, and Asia Pacific. For the student looking for nano-enlightenment on the cheap, they have great images and animations.

Meanwhile, on the other side of the world, the Asia-Pacific Nanotechnology Forum (www.apnf.org) fires up the discussion among Pacific Rim countries. The APNF's goal is to facilitate information flow between developers and investors throughout the region.

www.foresight.org

The Foresight Institute's Web site is a nonprofit organization with the mission of helping prepare society for future nanotechnologies. Since one of its main goals is promoting Foresight Institute founder Eric Drexler's vision of molecular manufacturing, it should come as no surprise that you can download Drexler's entire *Engines of Creation* text from the Web site. Nanodot, a spinoff of the Foresight Institute (www.nanodot.org), is a news-and-discussion Web site that provides general information about nanotechnology.

Other great sites

If you still have the energy after chasing down all this good info, check out these nanotechnology sites as well:

- **www.nsti.org:** The Nano Science and Technology Institute provides a good compilation of news broken down by subject (for example, microsystems, medical, materials, and so on). NSTI hosts the yearly Nanotech conference and trade show.

- **www.nanotech-now.com:** Nanotech Now provides general information on nanotech, including up-to-date news and a "Best of Nanotechnology" listing covering the last few years.

- **www.howstuffworks.com:** Although not specific to nanotech, this site breaks down difficult material into easy to understand visual explanations. It covers the basics of computers, electronics, science, and more.

Magazines

Yep, even the futuristic world of nanotechnology has a place for magazines. The online versions are easy to call up on-screen; the hard-copy versions you don't even have to plug in.

Technology Review

A journal published by the Massachusetts Institute of Technology, *Technology Review* aims to promote the understanding of emerging technology. Although not exclusively nanotech, it certainly picks some winners when describing which new technology makes it from the lab into the marketplace. This is the magazine subscription to get your nano-fan this Christmas. (For more info, check out their Web site at www.technologyreview.com.)

Small Times

Small Times is a nanotech magazine — and nothing *but* a nanotech magazine — providing comprehensive news coverage and searchable archives for those who want it all nanotech all the time. A lot of the main articles are free online but they also provide a bimonthly hard copy. (You can find out more at www.smalltimes.com.)

Science, Nature, and Nano Letters

These three journals are strictly meant for the hard-core nanotech enthusiasts — Nano Letters is strictly nanotechnology, whereas Science and Nature cover general science, including nanotech. The content is cutting-edge scientific articles — not for the technologically timid. However, if you're truly interested in the topic — and don't mind wading through the technical jargon — you can gain a priceless handle on the science behind nanotechnology. Speaking of price, these technical journals can have pretty high subscription rates, so check them out at your library and give them the once-over before you pull out your MasterCard.

✔ **Science:** www.sciencemag.org

✔ **Nature:** www.nature.com

✔ **Nano Letters:** Edited by two Movers and Shakers we mention in Chapter 15 (Paul Alivisatos and Charles Lieber), this magazine provides nano-exclusive articles. Take a look at their Web site (pubs.acs.org/journals/nalefd) to purchase either a subscription or individual articles.

Other great magazines

Go ahead. Reading is good for you.

✔ **Scientific American:** From a technical-coverage perspective, a step above *Technology Review* but still readable. Although general science, they have a section covering nanotech (www.sciam.com/nanotech).

✔ **National Geographic:** Will have a few articles in 2006 devoted to nanotech (www.nationalgeographic.com).

✔ **Wired, Popular Mechanics, and Popular Science:** All have had news and semi-technical nanotech articles. Keep them on your back burner: They're not always as up-to-date as the ones already listed.

- **Wired:** www.wired.com

- **Popular Mechanics:** www.popularmechanics.com

- **Popular Science:** www.popsci.com

Glossary

adenosine triphosphate (ATP): Organic molecule that stores energy in a biological cell.

amines: Organic compounds used as attachment points for molecular structures.

amphiphile: A molecule that has two distinct parts; a hydrophilic (water-loving) head and a hydrophobic (water-fearing) tail.

atom: Smallest particle of an element, composed of three types of charged particles: protons (positive), neutrons (neutral), and electrons (negative).

atomic force microscope (AFM): A scanning probe instrument that measures the atomic force acting on its tip as it moves along the surface of a sample.

band gap: The energy difference between the top of the valence band and the bottom of the *conduction band* in semiconductors and insulators. In insulators, the band gap is large — it requires a lot of energy to move *valence electrons* to the conduction band. In semiconductors, the band gap is not as large; it doesn't require as much energy to move electrons from the valence band to the conduction band. Adding impurities to the semiconductor (called *doping*) can change the band gap and the amount of energy needed to move electrons. In conductors (metals), the valence band and conduction band overlap, resulting in no band gap.

benzene: A ring of 6 carbon atoms, each with one hydrogen atom.

bioactive materials: Materials capable of interacting with living tissue.

bioavailability: The extent to which a drug successfully targets specific cells. Used in measuring the effectiveness of drug delivery.

biometrics: Identification based on personal features such as face recognition or fingerprint.

biomimetics: Applying systems found in nature to the design of engineering systems and modern technology. Velcro is an example of biomimetics: the plastic hooks and loops resemble plant burrs (hooks) that cling to animal fur and clothing (loops).

biosensor: A sensor that detects biological molecules such as proteins.

bottom-up fabrication: A construction process that works with the smallest units of a material first (in this context, atoms) and builds them up into the form of the final product. Compare *top-down fabrication*.

buckyball: Short for buckminsterfullerene; a molecule containing 60 carbon atoms in a soccer-ball orientation. Also known as *fullerene* or C_{60}.

buckypaper: A randomly oriented network of carbon nanotubes formed into a flat sheet.

cantilever: A solid beam allowed to oscillate at one end. Used in *atomic force microscopes (AFMs)*.

carbon nanotube: A graphite sheet rolled up into a tube.

catalyst: A substance that reduces the amount of energy required during a chemical reaction. Its presence increases the rate of reaction without the consuming the catalyst.

chirality: In the context of this book, chirality is the "twist" of a carbon nanotube. Twisting carbon nanotubes down the length of the tube gives them unique properties that depend upon the degree of the twist. (For example, specific twists make a nanotube either metallic or semiconducting.)

colloidal self-assembly: A process by which *colloids* assemble themselves into useful alignments; used in developing photonic crystals.

colloids: Very small particles (within the 1nm-to-1000nm range) that remain dispersed in a liquid for a long time. Their small size prevents them from being filtered easily or settled rapidly.

colorimetric sensors: Sensors that provide an indicator for quick macroscopic analysis by changing color.

composite: An engineered material composed of two or more components.

conduction band: The energy at which electrons can move freely through the material.

covalent bond: Atoms that bond sharing two electrons.

curing: Process of hardening. In this context, heat is added to a liquid polymer to harden it.

data mining: Sifting through large amounts of data, trying to find relationships and patterns within the information.

decoherence: Decoherence is the breakdown of quantum properties (superposition and entanglement) changing the behavior of the system from quantum mechanical to classical physics. This is usually the result when a quantum mechanical system interacts with its environment.

dendrimer: An artificial, polymer-based molecule that resembles a foam ball with tree sprigs shooting out of it in every direction. Contains a great number of voids between the sprigs, which allows it to carry drug molecules.

deoxyribonucleic acid (DNA): The nucleic acid that carries the genetic blueprint for all forms of cellular life.

diffraction: The spreading or bending of light as it passes by an object. An example of diffraction is using a prism to spread sunlight into a spectrum of color.

doping: Adding specific impurities ("dopants") to give a material desired properties, — as in the process that creates either n-type or p-type silicon.

electrochromatics: Materials that change color when energized by an electrical current.

electroluminescence: Converting electrical energy into light.

electron-beam lithography (EBL): Fabrication method that uses a tight beam of electrons to form nano-scale features on a substrate.

electro-osmosis: A method that uses an electric field to move liquids through a nano-channel. The sides of the nano-channel's wall are charged, allowing the liquid to slip through at a constant rate.

electrophoresis: A method of using an electric field to move particles through a nano-channel and separate them by size. The particles move at a rate inverse to their mass: the larger ones are slower than the smaller ones.

endocytosis: A process whereby cells absorb particles by enveloping them with the help of vesicles formed from the cell wall.

enigma: A mystery wrapped in a riddle. "Atomic interactions at the nano-scale are an enigma that is yet to be fully understood." To Rich, nanotechnology and women are both enigmas.

entanglement: Relationship in which the quantum states of two or more objects are always described with reference to each other, even if they're physically separate.

exocytosis: The removal of particles by enveloping them in a vesicle and releasing them outside the cell wall.

extreme ultraviolet (EUV): Light whose wavelengths are in the range of 10 to 200 nm, outside the higher end of the visible spectrum.

fabrication: Creating something physical. In the context of this book, the actual manufacture of computer processors.

fiber optics: Technology that uses light pulses through thin glass fibers at high speeds.

field-effect transistor: The most common type of transistor used in computer processors. It has a gate that controls whether it's a 1 or a 0.

fluorescence: A property of some molecules to absorb one wavelength of light and then emit light at a higher wavelength.

fullerene: A molecule containing 60 carbon atoms in a soccer-ball orientation. Also known as buckminsterfullerene, *buckyball*, or C_{60}.

functionalization: Attaching groups of molecules to a surface to serve a specific purpose.

graphite: A flat sheet of *benzene* rings attached together.

gray goo: Nanotech-disaster scenario in which myriads of self-replicating nano-assemblers make uncountable copies of themselves and consume the earth — "gray" because they're machines; "goo" because their small size makes them look like a thick liquid when taken together.

hemoglobin: Oxygen-carrying protein in blood cells.

holographic data-storage system (HDSS): High-capacity data storage, using pages of data rather than lines of data. This type of memory has both high capacity and high transfer rates.

hybridization: The process of joining two complementary strands of DNA together to form a double-stranded molecule.

hydrodynamic focusing: Using the properties of *laminar flow* to pinch and create a narrow stream of fluid at the micro- and nano-scale.

hydrophilic: "Water-loving" materials that are soluble in water. In a molecule, the part of the molecule that is attracted to water molecules.

hydrophobic: "Water-fearing" materials that do not dissolve in water. In a molecule, the part of the molecule that is repulsed by water molecules.

hysteresis: A property of magnetism: the magnetic effect doesn't disappear when an applied magnetic field is withdrawn.

impedance: The degree to which a wire resists the flow of electricity.

in vitro: Biological or medical experiments done outside the body, usually in a Petri dish.

in vivo: Biological or medical experiments done within a living subject.

lab-on-a-chip: Product that results from miniaturizing the processes of a lab (such as fluid analysis) into the space of a microchip.

laminar flow: Smooth and regular fluid flow. Opposite of turbulence.

laser: Acronym for "*l*ight *a*mplification through *s*timulated *e*mission of *r*adiation." An intense, powerful beam of light produced by this process is made up of nearly parallel waves.

liposome: A spherical *vesicle* composed of a *phospholipids* bilayer, used to deliver drugs or genetic material into a cell.

magnetic random-access memory (MRAM): Random-access memory that's based on magnets instead of capacitors. This type of memory is fast and non-volatile (that is, it doesn't disappear when you turn off the power), and uses less energy.

magnetic tunnel junction (MTJ): A type of magnetic random-access memory (MRAM).

metallofullerene: A metal atom caged in a *fullerene*.

metrology: The study of measurements.

micelles: Spherical micro-structures consisting of *amphiphiles*.

microelectromechanical system (MEMS): A mechanical system or machine that exists at the micro-level.

microfluidics: The study of the behavior of fluids at volumes thousands of times smaller than in a common droplet. Fluid at this level is very viscous; water moves like honey.

molecular electronics: Using organic molecules instead of silicon to make much smaller, faster, energy-stingier computer processors and memory components.

molecule: Two or more atoms chemically bonded together.

multiwalled carbon nanotubes (MWNT): Multiple carbon nanotubes within each other.

nano: Greek for "dwarf," meaning one billionth.

nanometer (nm): One billionth of a meter.

nanoshells: Gold-coated silica spheres which, when injected into the bloodstream, attach themselves to cancer cells. The nanoshells are then illuminated with a laser, giving off heat and killing the tumor cells.

nanotechnology: Technology development at the atomic and molecular range (1 nm to 100 nm) to create and use structures, devices, and systems that have novel properties because of their small size.

nanowire: Very small wires composed of either metals or semiconductors.

optical tweezers: A strongly focused laser beam used to grasp and move micro- and nano-size translucent particles.

organic molecules: Carbon-based molecules that make up the solid portions of living things, as well as certain materials such as plastics and oil.

organic surfaces: Surfaces that are non-metallic, such as skin, wood, or fabric.

oxidation: Chemically combining oxygen with another substance; fire and rust are two examples.

parallel processing: Simultaneous execution of the same task on multiple processors. Fast, nano-scale processors could make this technique possible on an unprecedented scale, as in the *quantum computer*.

pharmacogenetics: The study of how a patient's genetic make-up will affect his or her response to medicines.

phospholipids: Naturally occurring *amphiphiles* that make up human cell walls.

photolithography: A computer-processor fabrication technique that uses light to expose a photosensitive film, resulting in the needed pattern of circuits at a much smaller scale.

photon: A particle that is a packet of light.

photonic band gap: A *band gap* that corresponds to a specific wavelength of light used in photonic crystals. Photons that have this particular wavelength have to travel within this photonic band gap, restricted from the surrounding material. Useful for diverting light at the molecular level.

photonic crystal: A "light insulator" — materials that control how much (or what kind of) light is allowed to pass through the nanocrystal.

photonics: The science of manipulating photons.

photoresist: A substance that becomes soluble when exposed to light. Used in *photolithography*.

piezoelectric transducer (PZT): A material that expands and contracts according to the amount of electric current that travels through it.

plasma: A gas made of charged particles. An example of naturally occurring plasma is lightning. You may have seen plasma lamps, glass globes with sparks shooting around inside.

polymers: Plastic — large molecules made from many smaller molecules usually composed of carbon atoms bonded in long chains.

quantum computer: A computer that exploits the quantum mechanical nature of particles, such as electrons or atomic nuclei, to manipulate information as quantum sized bits (*qubit*). This quantum computer will be able to perform quick operations in parallel solving problems that can't be solved with today's computer (for example, factoring large numbers).

quantum cryptography: Cryptography scheme that relies on quantum mechanics to ensure accurate key exchange and prevent eavesdropping.

quantum dot: A semiconductor nanocrystal that exhibits quantum behavior in optical or electrical processes.

quantum mechanics: In physics, a theory that describes physical interactions between atoms more accurately than classical physics, often with results that seem strange from an everyday frame of reference.

quantum tunneling: A quantum-mechanical effect of transitioning through a state that classical physics would forbid. An analogy is throwing a ball at a wall and having it appear on the other side.

quantum: In atomic physics, a discrete and basic unit, similar to the way an individual electron is the basic unit of electricity. Plural form is *quanta*.

qubit: Quantum bit — smallest unit in quantum computing.

random-access memory (RAM): Memory storage that accesses data anywhere on the storage medium.

repeaters: In-line amplifiers that take fading light or electrical signals and resend them with more power.

respirocytes: Tiny mechanical spheres used to store and release oxygen directly within the bloodstream.

scanning electron microscope (SEM): Electron microscope that creates images of nanoscale features by bombarding the surface of a sample with a stream of electrons, scanning back and forth, and reading the reflected electrons as they bounce off the surface.

scanning probe microscope: An instrument that studies the properties of surfaces at the atomic level by scanning an atomically sharp probe over the sample. This produces an image of the sample's topography with atomic resolution.

scanning tunneling microscope (STM): The first scanning probe instrument — measures electrons tunneling between a scanning tip and a conducting surface.

Schottky barrier: Area of resistance to electrical conduction, occurring at the junction between the metal wires and the semiconductor in a computer processor.

self-assembled monolayers (SAMs): A single layer of atoms or molecules that has assembled itself under controlled conditions. This makes it possible to design surfaces at the molecular scale.

self-assembly: Process that creates the specific conditions under which atoms and molecules spontaneously arrange themselves into a final product. An example of self-assembly is the automatic arrangement of *phospholipids* into a cell wall.

semiconductor: Material that has more electrical conductivity than an insulator (which has no conductivity) but less than a conductor; it can be made to insulate or conduct electricity in patterns, as in a computer processor.

shape-memory alloy (SMA): A metal alloy that remembers its geometry. After it is deformed, it is heater to a specific temperature and regains its original geometry by itself.

single-electron transistor (SET): A *transistor* that switches between on and off (in computer terms, 1 and 0) by using a single electron — much smaller than a traditional transistor, which uses many electrons to switch.

single-walled carbon nanotube (SWNT): A *carbon nanotube* with one wall. Compare *multiwalled carbon nanotube*.

soft lithography: A process that uses polymers for molding and printing micro- and nano-structures. Pioneered by George Whitesides and used for *microfluidics* and its descendant, nanofluidics.

spectrometers: tools that reveal the composition of things by measuring the light absorbed or emitted by atoms or molecules.

spintronics: "Spin-based electronics" that exploits not only an electron's charge but also its spin.

sputter deposition: A method of creating a thin film of metal by sputtering fine particles onto a surface.

stent: An expandable wire mesh used to keep a blood vessel open.

strained silicon: New method of improving processor speed by stretching individual silicon atoms apart so electrons flow through a transistor faster with little resistance. Compare *superconductor*.

substrate: The supporting surface that serves as a base.

superconductor: A material through which electricity flows with zero resistance.

superlattice: A crystal formed of thin layers. A natural example is graphite.

superposition: When an object simultaneously possesses two or more values of a specified quantity. Useful in the development of *quantum computers*.

surface tension: The pull of a liquid into its most compact form to minimize the amount of energy used, keeping the surface area to a minimum.

surfactants: "Surface-active" molecules that reduce the surface tension between two liquids. Surfactants are used in many detergents as a dispersant between oil and water.

tetrapods: pyramid-shaped nanocrystals that resemble children's jacks.

top-down fabrication: A construction process in which we first work at the large scale and then cut away until we have a smaller product. This is similar to a sculptor cutting away at a block of marble producing the final product, a statue. Compare *bottom-up fabrication*.

transistor: A switch that determines whether a bit is a 1 or a 0.

uncertainty principle: In *quantum mechanics,* a principle made famous by Werner Heisenberg: Measuring one property in a quantum state will perturb another property. You can, for example, measure the position or momentum of an electron — but not both at once.

valence electrons: The electrons in the outermost shell of an atom. These electrons largely dictate the chemical reactions of the atom.

van der Waals: Weak electrostatic forces between atoms.

vesicles: *Micelles* with two layers: a reverse micelle surrounded by a regular micelle. Resembles the walls of biological cells.

viscosity: The measure of resistance of a fluid — its "thickness."

water window: A range of frequencies in the electromagnetic spectrum that are most easily transmitted through water, making them suitable for optical imaging (800-1300nm).

wavelength: In physics, the distance between one wave peak and the next in a transmitted wave of radiant energy. Typically measured in *nanometers*.

Index

• *F* •

• *G* •

Notes

SINESS, CAREERS & PERSONAL FINANCE

0-7645-5307-0

0-7645-5331-3 *†

Also available:
- Accounting For Dummies †
  0-7645-5314-3
- Business Plans Kit For Dummies †
  0-7645-5365-8
- Cover Letters For Dummies
  0-7645-5224-4
- Frugal Living For Dummies
  0-7645-5403-4
- Leadership For Dummies
  0-7645-5176-0
- Managing For Dummies
  0-7645-1771-6

- Marketing For Dummies
  0-7645-5600-2
- Personal Finance For Dummies *
  0-7645-2590-5
- Project Management For Dummies
  0-7645-5283-X
- Resumes For Dummies †
  0-7645-5471-9
- Selling For Dummies
  0-7645-5363-1
- Small Business Kit For Dummies *†
  0-7645-5093-4

ME & BUSINESS COMPUTER BASICS

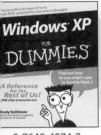

0-7645-4074-2

0-7645-3758-X

Also available:
- ACT! 6 For Dummies
  0-7645-2645-6
- iLife '04 All-in-One Desk Reference
  For Dummies
  0-7645-7347-0
- iPAQ For Dummies
  0-7645-6769-1
- Mac OS X Panther Timesaving
  Techniques For Dummies
  0-7645-5812-9
- Macs For Dummies
  0-7645-5656-8

- Microsoft Money 2004 For Dummies
  0-7645-4195-1
- Office 2003 All-in-One Desk Reference
  For Dummies
  0-7645-3883-7
- Outlook 2003 For Dummies
  0-7645-3759-8
- PCs For Dummies
  0-7645-4074-2
- TiVo For Dummies
  0-7645-6923-6
- Upgrading and Fixing PCs For Dummies
  0-7645-1665-5
- Windows XP Timesaving Techniques
  For Dummies
  0-7645-3748-2

OD, HOME, GARDEN, HOBBIES, MUSIC & PETS

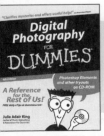

0-7645-5295-3

0-7645-5232-5

Also available:
- Bass Guitar For Dummies
  0-7645-2487-9
- Diabetes Cookbook For Dummies
  0-7645-5230-9
- Gardening For Dummies *
  0-7645-5130-2
- Guitar For Dummies
  0-7645-5106-X
- Holiday Decorating For Dummies
  0-7645-2570-0
- Home Improvement All-in-One
  For Dummies
  0-7645-5680-0

- Knitting For Dummies
  0-7645-5395-X
- Piano For Dummies
  0-7645-5105-1
- Puppies For Dummies
  0-7645-5255-4
- Scrapbooking For Dummies
  0-7645-7208-3
- Senior Dogs For Dummies
  0-7645-5818-8
- Singing For Dummies
  0-7645-2475-5
- 30-Minute Meals For Dummies
  0-7645-2589-1

TERNET & DIGITAL MEDIA

0-7645-1664-7

0-7645-6924-4

Also available:
- 2005 Online Shopping Directory
  For Dummies
  0-7645-7495-7
- CD & DVD Recording For Dummies
  0-7645-5956-7
- eBay For Dummies
  0-7645-5654-1
- Fighting Spam For Dummies
  0-7645-5965-6
- Genealogy Online For Dummies
  0-7645-5964-8
- Google For Dummies
  0-7645-4420-9

- Home Recording For Musicians
  For Dummies
  0-7645-1634-5
- The Internet For Dummies
  0-7645-4173-0
- iPod & iTunes For Dummies
  0-7645-7772-7
- Preventing Identity Theft For Dummies
  0-7645-7336-5
- Pro Tools All-in-One Desk Reference
  For Dummies
  0-7645-5714-9
- Roxio Easy Media Creator For Dummies
  0-7645-7131-1

eparate Canadian edition also available
eparate U.K. edition also available

ailable wherever books are sold. For more information or to order direct: U.S. customers visit www.dummies.com or call 1-877-762-2974.
. customers visit www.wileyeurope.com or call 0800 243407. Canadian customers visit www.wiley.ca or call 1-800-567-4797.

 **WILEY**

SPORTS, FITNESS, PARENTING, RELIGION & SPIRITUALITY

0-7645-5146-9

0-7645-5418-2

Also available:
- Adoption For Dummies
  0-7645-5488-3
- Basketball For Dummies
  0-7645-5248-1
- The Bible For Dummies
  0-7645-5296-1
- Buddhism For Dummies
  0-7645-5359-3
- Catholicism For Dummies
  0-7645-5391-7
- Hockey For Dummies
  0-7645-5228-7

- Judaism For Dummies
  0-7645-5299-6
- Martial Arts For Dummies
  0-7645-5358-5
- Pilates For Dummies
  0-7645-5397-6
- Religion For Dummies
  0-7645-5264-3
- Teaching Kids to Read For Dummie
  0-7645-4043-2
- Weight Training For Dummies
  0-7645-5168-X
- Yoga For Dummies
  0-7645-5117-5

TRAVEL

0-7645-5438-7

0-7645-5453-0

Also available:
- Alaska For Dummies
  0-7645-1761-9
- Arizona For Dummies
  0-7645-6938-4
- Cancún and the Yucatán For Dummies
  0-7645-2437-2
- Cruise Vacations For Dummies
  0-7645-6941-4
- Europe For Dummies
  0-7645-5456-5
- Ireland For Dummies
  0-7645-5455-7

- Las Vegas For Dummies
  0-7645-5448-4
- London For Dummies
  0-7645-4277-X
- New York City For Dummies
  0-7645-6945-7
- Paris For Dummies
  0-7645-5494-8
- RV Vacations For Dummies
  0-7645-5443-3
- Walt Disney World & Orlando For Dumm
  0-7645-6943-0

GRAPHICS, DESIGN & WEB DEVELOPMENT

0-7645-4345-8

0-7645-5589-8

Also available:
- Adobe Acrobat 6 PDF For Dummies
  0-7645-3760-1
- Building a Web Site For Dummies
  0-7645-7144-3
- Dreamweaver MX 2004 For Dummies
  0-7645-4342-3
- FrontPage 2003 For Dummies
  0-7645-3882-9
- HTML 4 For Dummies
  0-7645-1995-6
- Illustrator CS For Dummies
  0-7645-4084-X

- Macromedia Flash MX 2004 For Dumm
  0-7645-4358-X
- Photoshop 7 All-in-One Desk
  Reference For Dummies
  0-7645-1667-1
- Photoshop CS Timesaving Techniqu
  For Dummies
  0-7645-6782-9
- PHP 5 For Dummies
  0-7645-4166-8
- PowerPoint 2003 For Dummies
  0-7645-3908-6
- QuarkXPress 6 For Dummies
  0-7645-2593-X

NETWORKING, SECURITY, PROGRAMMING & DATABASES

0-7645-6852-3

0-7645-5784-X

Also available:
- A+ Certification For Dummies
  0-7645-4187-0
- Access 2003 All-in-One Desk
  Reference For Dummies
  0-7645-3988-4
- Beginning Programming For Dummies
  0-7645-4997-9
- C For Dummies
  0-7645-7068-4
- Firewalls For Dummies
  0-7645-4048-3
- Home Networking For Dummies
  0-7645-42796

- Network Security For Dummies
  0-7645-1679-5
- Networking For Dummies
  0-7645-1677-9
- TCP/IP For Dummies
  0-7645-1760-0
- VBA For Dummies
  0-7645-3989-2
- Wireless All In-One Desk Reference
  For Dummies
  0-7645-7496-5
- Wireless Home Networking For Dumm
  0-7645-3910-8

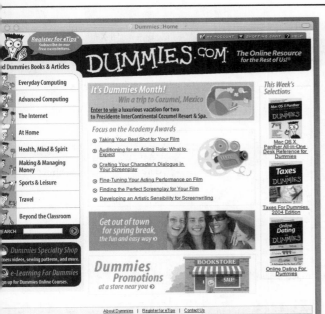